GENES, GIANTS, MONSTERS AND MEN

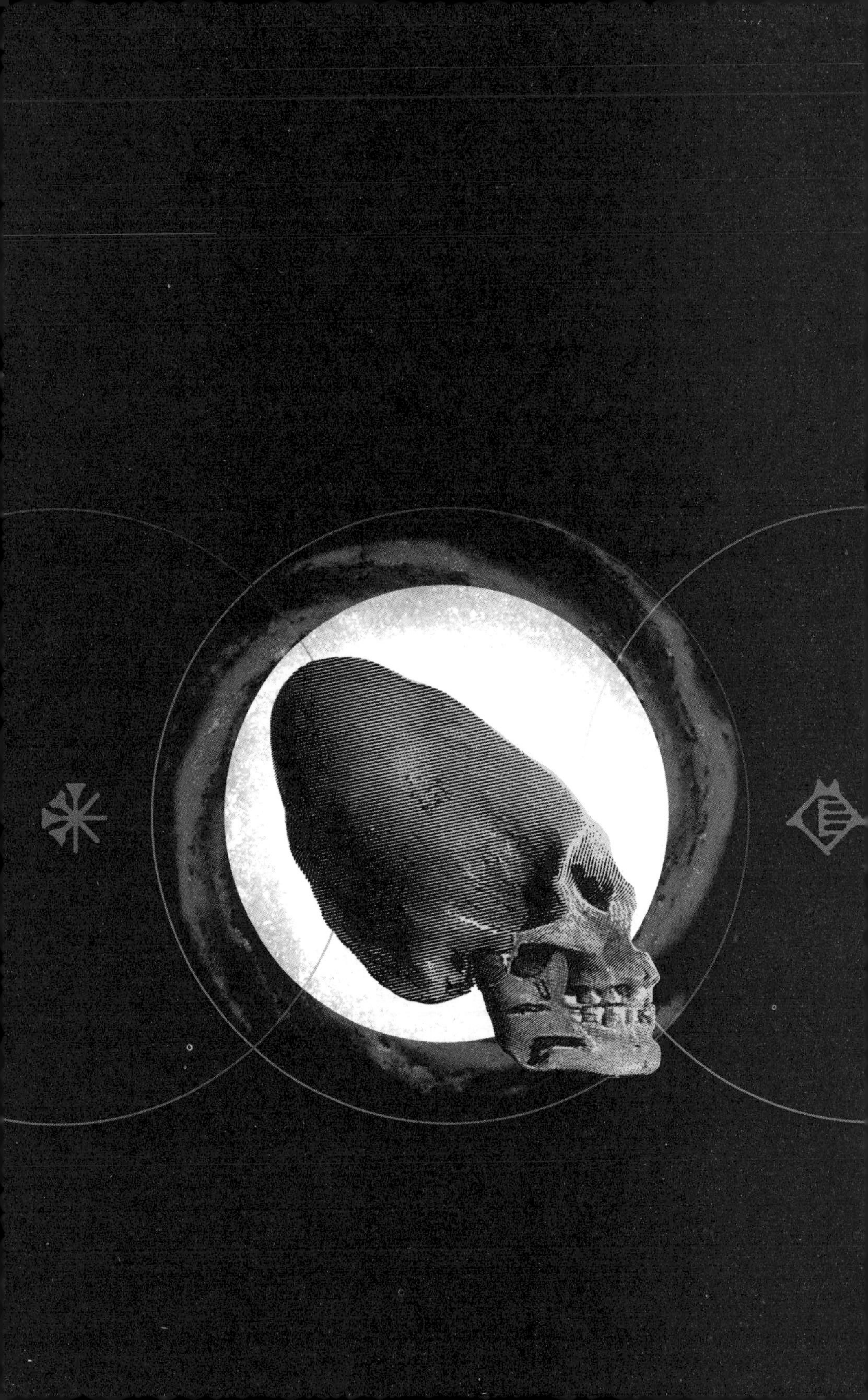

GENES, GIANTS, MONSTERS AND MEN

THE SURVIVING ELITES OF THE COSMIC WAR
AND THEIR HIDDEN AGENDA

JOSEPH P. FARRELL

FERAL HOUSE

"I think we're property.
I should say we belong to something:
That once upon a time, this earth was No-man's Land,
that other worlds explored and colonized here,
and fought among themselves for possession,
but that now it's owned by something:
That something owns this earth — all others warned off.

...

"I suspect...that all of this has been known, perhaps for ages,
to certain ones upon this earth, a cult or order,
members of which function like bellwethers to the rest of us,
or as superior slaves or overseers, directing us in accordance
with instructions received — from Somewhere else —
in our mysterious usefulness."

—Charles Fort, *The Book of the Damned*, 1919, p. 163

Genes, Giants, Monsters, and Men:
The Surviving Elites of the Cosmic War
and Their Hidden Agenda

A Feral House book
ISBN 978-1-93623-908-5

Feral House
1240 W. Sims Way Suite 124
Port Townsend WA 98368
www.FeralHouse.com
Book design by Jacob Covey

10 9 8 7 6 5

Above all, to

SCOTT DOUGLAS de HART:
A true master, adept, and poet of deep mysteries,
who crossed the Rubicon with me:
Anything I could say, any gratitude I could express,
are simply inadequate for you;
You are a true

And to

TRACY S. FISHER:
You are, and will always be, sorely missed.

ACKNOWLEDGEMENTS

Every book such as this one is the product of that intricate calculus of human interaction. It is the offspring of conversations between close friends with common interests and outlooks, of "conversations" with books written long ago by authors long forgotten, who yet in the immediacy of the media of information are ever present to us.

This book is no different. The authors from long ago and more recent times are encountered in its main text, but the conversations had with friends who have in their own way contributed to many of the insights of this book are a different matter, and these people must be mentioned. George Ann Hughes and I have known each other for about three years through her Internet broadcast, *The Byte Show.* What began as a professional relationship has morphed into a friendship where the free flow of ideas is always paramount, and many of her insights are recorded herein. Thank you, George Ann.

The same, likewise, must be said for my friend Richard C. Hoagland, who also over many emails and phone calls these past two years has made his own observations and contributions to this book, particularly in how space anomalies fit into the larger picture. I hope I have done some justice to his thoughts here. Thank you, Richard.

Above all there has been one friend, Dr. Scott D. deHart, whose constancy as a friend in some of the darkest periods of my life has been unwavering, and whose conversations over this last third of my life that I have had the honor and privilege of knowing him, have so mightily contributed to the ideas presented herein that in truth, he is a co-author and equal contributor to it. Many of the ideas, particularly in the last two parts of the book, are reflective of conversations we've had over the past decade. For him and for these conversations and his friendship, and in many cases mentorship, no words of gratitude suffice, but nonetheless I say it anyway: thank you, Scott.

Joseph P. Farrell
Spearfish, South Dakota
2010

Introduction

∴

HOW DOES ONE MAKE SENSE of ancient texts, modern genetics, legends of monsters and giants, fossils, the origins of man, and the attempts of some religions, to obscure all of it? How does one *synthesize* it all? Is it even possible to speak in one breath of giants, monsters, ancient texts and evolution?

Asking these questions highlights the problem, and the problem is the texts, what *we* mean by them, and what we think *they* mean.

Pretty much everyone with half a brain, and who has not been lobotomized by the American "education" system or subjected to the psychedelic drugs and mind-numbing electroshock of its dull university curricula, even duller professors, and to its "textbooks" that contain no primary texts, is agreed that *something* is wrong with our standard model of history, particularly the farther back one goes. Only a university academic, for example, could believe that humanity was around for about 150,000 years (if one is to believe the geneticists), and doing nothing but "hunting and gathering," and then all of a sudden, and for no explicable reason, decided to invent civilizations such as Sumer (and all its Mesopotamian offshoots) and Egypt out of whole cloth, undertake monumental ziggurat or pyramid construction, invent calendars, agriculture, wheels, writing, mathematics, music, astronomy, banking, and maybe even electricity, as evidenced by the Baghdad Battery.

Consequently, when it comes right down to it, we have a choice between fairy tales, or, if one prefers, between mythologies or dogmas.

We can believe the hypnotic incantations of biology, history, and anthropology professors waving their wands, and producing cute animated videos of the primordial soup gradually evolving into complex organic life (fish), whose fins gradually morph into limbs as they crawl out of the soup onto the sand (reptiles) and, as the computer-generated animation proceeds, gradually morphing into a veritable cornucopia of evolutionary progress over "billions of years" (insert Carl Sagan voice here).

Or we can begin to take seriously what those ancient societies — that suddenly sprang up out of whole cloth and almost *ex nihilo* — said happened. There the picture is equally disconcerting, for one is almost immediately confronted by wild tales of giants, monsters, men and even — if one reads the texts a certain way — with genetic manipulation and extraterrestrials, and a cast of characters that range the whole spectrum from nobility to a genocidal disposition and behavior that would make a Stalin or a Hitler seem like paragons of virtue, restraint, and compassion.

Indeed, the comparison with Stalin and Hitler is an apt one, for if one reads certain ancient texts from Sumeria or India, one encounters over and over such characters, whether gods or men, engaged in a titanic struggle for power that ultimately resulted in a "cosmic war," and its aftermath. Yet, if one is to believe the ancient stories, the monsters and giants and manipulations remained after that war was over, scattering themselves along with some of the technology by which it was fought, yet doing so with no consistency. For example, one finds both the Sumerian *Anunnaki* and the Hebrew *Nephilim* siring chimerical and gigantic offspring with human women, and yet the context in which the story is told — Sumerian "polytheism" versus Hebrew "monotheism" — changes drastically, and implies a hidden agenda (or agendas) behind each version of the story.

This book looks at things from the standpoint of the *second* fairy tale; that is, it *assumes* that a very ancient "cosmic war" was fought within our own solar system, by a civilization with interplanetary extent, and which was based on this planet, the Moon, Mars, and various satellites of the gas giants. As I have detailed elsewhere, that war was so destructive that the civilization waging it was nearly wiped out, leaving its surviving elites to pick up the pieces as best as they could, and begin the long, slow climb back up to a similar pitch of scientific and technological development.[1] The course of that war, and its aftermath, was populated by a bizarre pantheon of gods, of giants — the chimerical offspring of the "gods" and man — and by even more chimerical monsters, and of course, by humans. These, and the deeper

1 Q.v. my *The Cosmic War: Interplanetary Warfare, Modern Physics, and Ancient Texts* (Kempton, Illinois: Adventures Unlimited Press, 2007).

implied issues of a sophisticated genetic science in existence in very ancient times, constitute the obvious motifs of this book. *They are the portals by which we will enter into the central theme of this book, which is to acquire a preliminary understanding of the agendas of those elites that survived the cosmic war and catastrophe, and to trace a broad chronology of their activities.*

Thus, our focus is both on the elites and their agendas in the aftermath of that war, and with this, certain methodological assumptions are implicit. We assume, for example, that:

1) Elites *did* survive, both in pockets scattered over the surface of the earth, and possibly in pockets *altogether off this planet on other celestial bodies, wherever those might be;*
2) Like all elites, these had certain agendas, some of which it had to put into place immediately to ensure their, and humanity's, survival, and some of which were of a more long-term nature, such as moving civilization by progressive steps back to a similar pitch of technological and scientific achievement as it had before the war. This would require a global extent to such a civilization, and all the trappings of civilized society, including, especially, agriculture and commerce.
   a) For the conduct of commerce, especially, an accurate system of weights and measures would be required, and this would have to be of fairly universal extent, and one moreover that was capable of simple and accurate reproduction anywhere on the planet. It is a typical "engineer's optimalization" problem: how would one do so within the constraints posed by relatively simple and primitive tools that were probably the only tools left after the Cosmic War had all but blasted apart the infrastructure of whatever high civilization as once existed, and do so by deriving a method that would be extremely accurate, anywhere on the earth? The most obvious and ready-to-hand systems available would have been those based upon the relative constancy of astronomical and geodetic phenomena. As we shall discover in the main text, there is ample evidence to suggest that the earliest systems of measure were indeed based on these foundations, and once spread, the more long-term goal of civilization building began in sudden earnest.
   b) Similarly, as tools of cohesion, conquest, and a considerable degree of obfuscation, religions were promoted by the very

same elites, and as commerce and contact between civilizations grew — often fostered by the very same elites — so too did religious agendas change, often violently, but just as often subtly. And here, suggestively, there were remarkable recreations of a lost technology by which oracles and revelations could, to a certain extent, be staged and coordinated. This is *not* a claim or assertion that all religions from that postbellum period, and which claim an oracular or revelatory foundation, are the products of an "oracular technology," but only that some of them *could* be.

3) The careful reader will have noticed yet another implicit assumption, and that is that in order for any such post-Cosmic War elites to function with such long-range goals, they must of necessity operate not only in a coordinated fashion, but in a fashion *continuous* over time and throughout history.[2] They will, in short, have their modern descendants, and these too will evidence their own agendas, and while we will not often touch upon them in this book, we will encounter them at work in the first three chapters. Tracing the memberships and lineaments of those elites in ancient times is, of course, next to impossible if one hopes to dig up cuneiform "lists of secret members." But one may approach the task by looking at the policies and practices of the known elites of those times, and there the picture becomes very suggestive if not completely persuasive: commerce, religion, and civilization itself were in the hands of an astronomical-astrological priesthood — complete with all the homosexual and heterosexual temple prostitution that accompanied it — and that elite worked with cold calculation behind the scenes to raise up, and to destroy, whole civilizations in pursuit of its ultimate goal: the reestablishment of a global civilization. *That* goal had to be reached, if humanity was ever to return to the stars, and reestablish contact with part of that pre-Cosmic War humanity — its "cousins" — with whom it had lost touch.

This list of implicit methodological assumptions may be summarized by a series of simple questions that will be touched upon in this book: *why does such an agenda (or agendas) even exist, who is behind it, and what is its purpose?*

---

2 For the question of the continuity of such esoteric groups through history — though a much later period of history than is being examined here — see my *The Philosophers' Stone: Alchemy and the Secret Research for Exotic Matter* (Port Townsend, Washington: Feral House, 2009), pp. 27–29.

What do the ancient stories and modern science say or imply about this agenda? Can what they say be harmonized? Perhaps last, but by no means least, *what is the connection between this agenda and the "cosmic war of the gods" that preceded it?*

Such questions also highlight the technological themes of this book, for to suggest that elites existed with an agenda to manipulate humanity *en masse* via genetics, technologies, and religion, and to do so over a great period of time in order to reach a millennia-distant goal suggests at least rudimentary technologies and techniques had to be in place, and accordingly, we will look for evidence of these things in the archaeological record and the texts themselves. Additionally, I have already assumed the existence of a Very High Civilization that fought a cosmic interplanetary war, and even blew up a planet in our solar system, in my previous book *The Cosmic War.* Moreover, as I have argued in *Babylon's Banksters,* there is abundant evidence that these elites were fully international in extent, that they manipulated both finance and religion themselves, *and had a rudimentary technology of communication by which to do so.*[3] The implications of these assertions are rather obvious, for a civilization capable of blowing up entire planets implies a technological sophistication. But such sophistication also implies a similar nadir of development once existed in terms of social engineering and the technologies to manipulate man himself, that is to say, to manipulate his *mind,* his *brain,* and his very life processes — his *DNA.*

I assume, therefore, in the main body of this work that some of these technologies, and particularly those of brain-mind manipulation, possibly survived for a brief period after that Cosmic War, or that they were at least reconstructed in rudimentary form. As such, the mere *existence* of the *possibility* of the use of such technologies in ancient times reveals not only the possible agendas of those elites, but also raises the vexing possibility that the great "revelations," beginning with Abram/Abraham and Moses, and the religions founded upon them, were nothing but the productions of a technology, and to serve an agenda more or less hidden from those being manipulated. As will be seen in the main body of this work, the reading of ancient texts in the light of modern technology inevitably poses a significant problem for religious apologetics of the revealed religions. The task of this book, however, lies not in apologetics, but in synoptics; the task is not to solve the apologetic dilemma but as thoroughly as possible to pose the questions and the problems.[4]

---

3 Joseph P. Farrell, *Babylon's Banksters: The Alchemy of High Finance, Deep Physics, and Ancient Religion* (Port Townsend, Washington: Feral House, 2010), pp. 251–264.

4 Those with an apologetical turn of mind will note that the nature of defining revelation itself is greatly tightened by posing such questions: how *would* one distinguish the "voice or

These types of questions throw open the whole notion of what a "text" is, and how it might be interpreted. As I wrote in the predecessor to this book, *The Cosmic War*:

> Careful consideration of the questions outlined above, and of the parameters of the "interplanetary war scenario" itself, will also reveal the *types of evidence* to be considered...: (1) physics, (2) the material evidences of anomalous artifacts, (3) evidences and mechanisms of planetary destruction, and finally, and by no means the least important, (5) textual and "legendary" evidence from texts, oral myths and traditions, and physical monuments and ancient glyphs. "Text" in other words is understood in this book in the broadest sense, as being inclusive of all these things.[5]

Our focus in this book is obviously on the last of these five points, on both the "text" in the proper sense of actual written artifacts, oral myths and traditions, and upon the physical monuments themselves, and upon *how they might illuminate those hidden agendas and their purposes.*

But now a new consideration enters the picture, and that is the meaning of "monument" or "artifact" within the context of the above methodological assumptions about "texts," for it will be evident upon a little consideration, that DNA itself is a monument, an artifact, a text that can be decoded via the decryption techniques of modern genetics. This step, once embarked upon, yields curious anomalies of its own, for it leads to even deeper connections to other, more occulted traditions and texts; it also leads to deep connections between the physics of those anomalous monuments, the manipulations of DNA recorded in those ancient stories, and the texts themselves. There are, so to speak, "codes" within the "code," and some of these codes were perhaps known in very ancient times, a relic, a legacy of that sophisticated scientific culture that blew itself apart. There is, as we shall see, a suggestive and direct link between human DNA and the physical medium itself, and it comes from an unsuspected time and place.

As such, this book, like all my books on ancient lore, esoteric doctrine, and science, is a book of *high speculation*. It is argued speculation, to be sure, but it remains speculation nonetheless. In this case, however, there is an additional factor, and that is its cursory nature, for to explore any one of the

---

command of God" from the production of a technology and a command to go slaughter whole peoples for whatever reason?

5 Joseph P. Farrell, *The Cosmic War: Interplanetary Warfare, Modern Physics, and Ancient Texts* (Kempton, Illinois: Adventures Unlimited Press, 2007), p. v.

topics outlined herein in any fashion even close to thorough would require a book for each topic. Consequently, this book is but synoptic panoramic overview. It is an exploratory essay on how such post-Cosmic War matters *might* profitably be viewed; it is therefore perforce not a dissertation on all possible ways to view it.

Joseph P. Farrell
Spearfish, South Dakota
2010

TABLE OF CONTENTS

Epigraph v
Dedication vii
Acknowledgements ix
Introduction xi

PART ONE 3
POST-BELLUM AFTERMATH:
AN ANTHOLOGY OF ANOMALIES, ELITES, AND AGENDAS

1. KOLDEWEY'S CONUNDRUM AND DELITZSCH'S DILEMMA 5
A. Koldewey's Conundrum: The Sirrus 6
B. Delitzsch's Dilemma: Babel und Bible 14
1. The Cuneiform Tablets and an Out-of-Place Name for God 20
2. The Documentary Hypothesis: Astruc to DeWette 21
3. The Documentary Hypothesis: Hupfeld's "Copernican Revolution" 24
4. The Documentary Hypothesis: Karl Heinrich Graf and Julius Wellhausen 25
5. The Critical Suspicion of an Agenda 26
6. The Suggestions of an Agenda at Work with the Critics: Weishaupt's Strange Comment and a Hidden Illuminist Role in Early Old Testament Criticism? 27
7. The Explosive Thunderclap of Delitzsch's Dilemma 28

2. "MARDUK REMEASURED THE STRUCTURE OF THE DEEP": THE MEGALITHIC YARD AND THE SUMERIAN REFORM 31
A. An Oxford Professor Overturns the Standard Model: Alexander Thom, His Work, and its Implications 33
1. Thom's Megalithic Yard and the Expansions of Knight and Butler 35
2. The Methods 37
a. Thom's "Ancient Bureau of Standards Theory" 37

3. Celestial Geometries and the Pendulum Method 38
4. Beautiful Numbers: The 366-, 365-, and 360-Degree Systems 41
5. The Next Step: Measures of Weight and Volume 44
6. Ancient and Megalithic Anticipations of the Imperial and Metric Systems 45
B. The Hidden Elite and the Cosmic War Scenario 46
1. The Ancient Elite: Astronomy, Finance, and the God of Corn versus the God of Debt 46
2. The Masonic Elite and the Lore of "Very High Antiquity" 49
C. The Cosmic War: Marduk Measured the Structure of the Deep 53
1. The Suggestions of a Deeper Physics 56
2. The "Sumerian" Mysteries of Deep Space 58
3. A Chronological Conundrum 63
D. Conclusions 63

3. The Oracular Technologies of Revelation: Mind Manipulators, Torsion Temples, and Religion Revealers 67
A. David Koresh Hears the Voice of God... or Was it Just Charlton Heston? 69
B. The Mind Manipulators: Mind Manipulation Technologies 74
1. The Alien Abduction Scenario and the Moral Disconnect 74
2. The False Dialectic, Occam's Razor, and Their Implications 75
3. Electronic Methods of Mind Manipulation 77
a. Electromagnetic Fields, Implants, and Combinational Approaches 78
b. The Remote Induction of Trances and "Hearing Voices" 79
c. Electronic Dissolution of Memory: Missing Time, and Missing History 81
d. "Abductions" and "Revelations": a Common Method 83
4. The Beat Frequency of the Brain 84
5. The Beat Frequency, and Out-of-the-Body Experiences 85
6. The Remote Induction of Trances, Emotions, Specific Information, and Remote "Telepathy" 85
7. Electromagnetic Alteration of DNA 86
8. Section Summary 86
C. The Torsion Temples of Antiquity: A Primitive Technology of Communication and Special Revelation 87
D. Religion Revealers: The O'Briens' Technological Consideration of the revelation of YHWH and the Torah 90
1. Yahweh as One of the Shining Ones 93
2. A Genetic Agenda? 93
3. Yahweh's "Pillar of Fire": The Technological Component 94
4. The Threats Begin: The Carrot, the Stick, and Other Techniques 96

a. Cowing Through Technology and Public Executions 102
b. An Indefensible Act: Impalings 106
5. An Inhuman Face? 107
6. The Tabernacle: Yahweh's Mobile Palace 108
7. The Urim and Thummim: More Technology? 109
8. The Shining Ones and the Possible Agendas 111
9. "Reverse Depatterning," Psychic Driving, and the Stockholm Syndrome 112

4. Elites With Agendas: Conclusions to Part One 117

PART TWO 123
Genes and Giants,
or,
If "It" is in the Genes, Then What is "It"?

5. The Genome Wars: Modern and Mesopotamian 125
A. An Overview of the Modern Genome War: The Race Between the Human Genome Project and Craig Venter's Celera Corporation 126
B. Technicalities and Legalities 130
1. The Technology: Sequencers 130
2. The Technology: Super-Computers 133
3. The Technique: Computer Algorithms 133
4. The Legal Implications 136
C. The Potentialities of Genetic Engineering 137
D. The "Mesopotamian" Genome War: The O'Briens Again 138
1. The Anunnaki and the Engineering of Man 144
2. The O'Briens on the Technological Indications 147
3. Verification: Genetics, Space, and Skeletons 155
4. Whose Agenda, Public, or Private? 156

6. A Connection of Miscellanies: The Codes within the Code, and the "Archaeology Conspiracy" 161
A. The Codes Within the Code 161
B. Back to the Tablets of Destinies 164
C. Giants and the "Archaeology Conspiracy" 169
1. The *New York Times* Reports the Discovery of Giant Remains 169
2. Egypt in Arizona 171
3. Hoax or Cover-up? 178
a. Destroyed and Suppressed Evidence: Curious NASA Parallels 182

b. The Smithsonian and the Suppression of the Alaskan Giants 183
4. Archaeology-gate: Cremo, Thompson, and the Antiquity of Man 184
5. An Aside: Gilgamesh Discovered — Speculative Implications 186
D. Conclusions: Taking Stock Thus Far 187

PART THREE 191
MONSTERS AND MEN

7. THE RETURN OF THE SIRRUSH: GREEKS, INDIANS, GIANTS, AND MONSTERS 193
A. Greeks, Giants, Monsters, and War 199
1. The Gigantomachy, or The War Against the Giants 199
2. The Griffin 201
3. Mayor's Interpretive Paradigm 201
B. Indians, Giants, Monsters, and War 202
1. The Age and War of the Giants and Monsters 202
2. The Consistency Native American Explanations 207
3. Mayor's Explanation 207
4. The Ancient Traditions and the Alternative Explanation 208

8. A MEMORY OF MAN PAST: GENETIC CLANS, ARCHAEOLOGICAL ANOMALIES, EVOLUTIONARY ENIGMAS, AND SPECULATIVE SOLUTIONS 213
A. Mitochondrial Eve and Her Seven European Daughters 214
1 Early Attempts to Distinguish Racial Groups by Blood-Typing 214
2. The Basques, Rh Positive and Rh Negative Blood, and "The Problem of Europe" 215
3. Mitochondrial DNA and the Y Chromosome: Mitochondrial "Eve" and Y Chromosomic "Adam" 219
a. Mitochondrial DNA and the Seven Clans 219
b. The Y Chromosome and the "Ten Fathers" 220
4. The Seven Mothers of Europe and Their Clans 221
a. Ursula's Clan 221
b. Xenia's Clan 222
c. Helena's Clan 222
d. Velda's Clan 222
e. Tara's Clan 222
f. Katrine's Clan 222
g. Jasmine's Clan 223
h. The Deeper Ancestry, the Beginnings of a Problem, and Some Beginning Speculations 223
B. Evolutionary Chronology of the Origins of Man and the Chronological Problem 225

1. Evolutionary Chronology of Human Origins and Proto-Humans 225
2. The Cremo-Thompson Archaeological Anomalies and Genetic Antiquity Problems 227
3. The Chronology of the Cosmic War and the Ancient Texts 231
C. Chronological Resolutions and Agendas: Some Speculations 233

Bibliography 243

GENES, GIANTS, MONSTERS AND MEN

I.

Post-Bellum Aftermath:

An Anthology of Anomalies

"When we speak of suppression of evidence, we are not referring to scientific conspirators carrying out a satanic plot to deceive the public. Instead, we are talking about an ongoing social process of knowledge filtration that appears quite innocuous but has a substantial cumulative effect. Certain categories of evidence simply disappear from view, in our opinion unjustifiably."

—Michael A. Cremo, "Introduction and Acknowledgements," from *The Hidden History of the Human Race,* p. xvii.

"In addition to the general process of knowledge filtration, there also appear to be cases of more direct suppression."

—Michael A. Cremo, "Introduction and Acknowledgements," from *The Hidden History of the Human Race,* p. xviii.

One

Koldewey's Conundrum and Delitzsch's Dilemma

∴

"At first this fabulous creature was classed along with the winged, human-headed bulls and other fabulous monsters of Babylonian mythology, but profound researches gradually forced the professor to quite a different conclusion."
—Ivan T. Sanderson[1]

ROBERT KOLDEWEY, famous German architect and "amateur" archaeologist, faced a problem. A *big* problem. In the intellectual world of the nineteenth century, the myth that all ancient myths were nothing *but* myths was quickly collapsing. Von Schliemann would prove that ancient Troy, far from being a figment of Homer's overactive and quite epic Hellenic imagination, actually existed, for he was the one who, using clues from Homer's "myth," actually dug it up. Whoops. Sorry, academia. Wrong again.

Koldewey also entered this typically German quest to verify the reality of ancient myths not only by unearthing Babylon from her sandy tomb, but the actual fabled "hanging gardens,"[2] one of the seven wonders of the ancient world, and the equally impressive Ishtar Gate of Babylon. He was one of the principal architects, in fact, of what would become something of an archaeological obsession with the region for the Germans, and they've been there ever since, scratching in the sands of Mesopotamia for clues to the actual history of mankind. And that was

1 Ivan T. Sanderson, *More "Things"* (Adventures Unlimited Press, 2007), p. 21.

2 Many modern scholars and archaeologists dispute Koldewey's conclusions and doubt he found the hanging gardens. He did, unquestionably, find Babylon and the Ishtar Gate.

the problem, for the deeper they dug, the stranger that picture became. And in Koldewey's case, the problem was even more acute, for the problem *was* a picture.

The problem was a picture, or to be more precise, the ideas he was entertaining about that picture, for it was one thing to maintain Troy and Babylon really existed, but *this?* Could it be? And if so, what would the academic world think? Had he been under the desert sun too long? Had he a touch of *Wahnsinn*? Was he perhaps *ein bisschen Verrükt?* He surely must have wondered those things himself, given the thoughts he was conceiving, not to mention the fact that he was actually thinking about *publishing* those thoughts. But the insanity of World War I still raged... perhaps no one would notice (until it was too late) if he just snuck a most unorthodox, nonacademic "idea" into an otherwise serious scholarly and archaeological study. After all, he needn't comment on its implications, which were many and profound. He could leave commentary to others. All *he* had to do was "sneak it in," point the way, *hint* at those wide and profound implications.

And that's exactly what he did in a book published in Leipzig in 1918.

The book was innocently entitled *Das Ischtar-Tor in Babylon, The Ishtar Gate in Babylon.* And like the Ishtar Gate itself, Koldewey's book will be our gate into a very epic, and very Babylonian, problem.

A. Koldewey's Conundrum: The Sirrush

The picture, or rather, bas-relief, that was causing the good Professor Koldewey such grief was this, the middle animal on either side of the Ishtar Gate of Babylon, the reconstruction of which is shown below:

The Reconstructed Ishtar Gate of Babylon

And a close-up of the left side will reveal the problem:

Close-up of the Ishtar Gate Animal Reliefs

Note the top and bottom reliefs, like so many other reliefs in Babylonian and Assyrian artwork, are of fairly conventional-looking cattle or other very ordinary animals. But these were not Professor Koldewey's problem. The problem is the *middle* relief, appearing as it does between two very normal-looking bulls.

A closer look at that middle relief is in order:

The Creature from Babylon: The Sirrush

While the head of this creature — whatever it is — is obscured somewhat in the photo, the feet alone should tell us, as they told Koldewey, that "we have a problem," for the front "paws" look somewhat like the paws of a large feline, while the rear "claws" look everything like the claws of some gigantic bird.

As if that were not enough, there is a long "spiraling" tail...

Spiraling Tail of "the Creature from Babylon"

...a long thin body that appears to be feathered or scaled...

The Long Scaled or Feathered Body of "the Creature from Babylon"

...and topping it all off was the head of a dragon or serpent of some sort:

Dragon's Head of "the Creature from Babylon"

However, the problem for Koldewey (and everyone else since, as we shall see shortly) was not that the Babylonians had given full freedom to their artistic flights of fancy; the problem was that they apparently had *not*, for the creature, known as a *"Sirrush,"* appeared right in the middle of other creatures known as aurochs that were self-evidently real, and though they are now extinct as well, they were not extinct in Babylonian times. The problem was the very real *context* in which the otherwise fantastic and bizarre "Sirrush" appeared. But that wasn't the *only* problem.

Koldewey wrote:

A creation of another, essentially different type confronts us in the "dragon." This is the *sirrush* of legend, or as it is often referred to today, the *Mus-rushu*, which Delitzsch renders as "splendid serpent."

The slender body, the wavy-lined tail, the similarly steep, solemn slender neck with its small scale-covered head... stands out better in color reproduction. The scaly attire shows itself on the hind legs downward to the middle of the shins. One observes larger diagonal scales on the abdomen. The forelegs resemble those of a long-legged type of cat, perhaps a panther. The hind feet are those of a bird of prey.... On the end of the tail one can observe a curved quill, as in a scorpion. The head is entirely that of a snake with a closed mouth from which a forked tongue protrudes. It also bears a large upright, prominent horn from which an appendage spirals or curls out.... Behind the "whiskers" a tuft of three locks of hair falls, pictured as three long spiraling locks....

This strange animal, with the above-enumerated features, as per Jastrow's picture portfolio of the religion of Babylon and Assyria, was found in the oldest Babylonian art and preserved these features unchanged for millennia. *Thus one may not say that it is a fantastic production, a chimerical picture of Babylonian-Assyrian art.*[3]

---

3 Robert Koldewey, *Das Ischtar-Tor,* Ausgrabungen der Deutschen Orient-Gesellschaft in Babylon (Leipzig: J.C. Hinrichs'sche Buchhandlung, 1918), pp. 27–29, my translation from the German, emphasis added. The complete text of the German cited above (including passages elided in the citation above) is as follows:

"Eine Schöpfung wesentlich anderer Art begegnet uns in dem Drachen. Es ist der "Sir-russu" der Inschriften oder, wie heute vielfach gelesen wird: "Mus-russu", von Delitzsch durch "Prachtschlange: wiedergegeben.

"Der schlanke Leib, wie der in Wellenlinie emporgerechte Schweif, und der ebenfalls steil getragen dünne Hals mit dem kleinen Kopf ist mit Schuppen bedeckt, die einen dachförmigen Querschnitt haben, was in der Photographie nach dem night emaillierten Exemplar...besser hervortritt als in der Wiedergabe der farbigen. Das Schuppenkleid zieht sich an den Hinterbeinen bis zur Mitte des Unterschenkels herab. Am Bauch bemerkt man die grösseren Querschuppen. Die Vorderbeing ähneln denen einer hochbeinigen Katzenart, etwa einers Panthers. Die Hinterfübe sind die eines Raubvogels, mit ausnahme des Tarsalgelenks, dessen Bildung noch dem Vierfübler zugerechnet ist. Am Schweifende bemerkt man einen kleinen nach oben gebogenen Stachel, wie ign der Skorpion hat. Der Kopf ist im ganzen der einer Schlange, bei welcher auch bei geschlossenem Maul die gespaltene Zunge hervortritt. Er trägt ein grobes, aufrechtes, gerades Horn, von dessen Wurzel aus ein Hautanhang sich spiralig nach hinten emporringelt. Beide Bildungen sind, wie schon oben bemerkt, paarig aufzufassen. Hinter den Backen fällt ein Büschel von drei Haarlocken herab, und eine aus drei langen, ebenfalls spiraligen Locken gebildete Mähne zieht sich vom Kopf bis zur Schultergegend herab.

"Dieses merkwürdige Tier, als das des Gottes Marduk un zugleich des Nebo, tritt mit den hier aufgezählten Haupteigenschften schon in der ältesten babylonischen Kunst auf Jastrow, Bildermappe zur Religion Babyloniens und Assyriens, und hat sich Jahrtausende hindurch fast unverändert erhalten.

In other words, one had a creature with the forelegs of a great cat, the hind legs of a bird, with a curving tail with what appeared to be a scorpion's sting, a long scaly body, a snake's head, out of which grew a horn! And this creature appeared in the artwork of the region with *amazing* consistency through the millennia, and in the context of other very *real* creatures, one of which was the now-extinct *aurochs* (about which more in a moment). It could not be, Koldewey concluded, merely the chimerical production of a fevered Mesopotamian artistic imagination, for in cases where such mythological creatures were encountered in Babylonian art, these showed a great deal of change over time; the *sirrush* did *not.*

Koldewey attempted to rationalize the creature's strange appearance by various comparisons to the features of known dinosaurs, and concluded, somewhat less than convincingly, that "When one finds a picture such as our *sirrush* in nature, one must reckon it as belonging to the order of dinosaurs and indeed of the sub-order of ornithopods."[4] However, one would be hard-pressed to find dinosaurs with the forelegs of a cat, the hind legs of birds of prey, with spiraling tails and scorpion's quills, and snake's heads growing horns, all in one fantastic creature. Koldewey proposed one dinosaur, the iguanodon, which did indeed have hind feet similar to a bird, as being a close match to the *sirrush.*[5] But that does not really make the dilemma any more palatable, since that would mean that *long* after dinosaurs were supposed to be extinct according to standard evolutionary theory, the ancient Babylonians were depicting them in the clear context of other very real, and very *living,* creatures, the aurochs.

To make matters very much worse, it even appeared to Koldewey that the *sirrush* might have been the basis behind at least one biblical story, that recounted in the Greek versions of the book of Daniel, and known as *Bel and the Dragon:*

> And in that same place there was a great dragon, which they of Babylon worshipped. And the king said unto Daniel, Wilt thou also say that this is of brass? lo, he liveth, he eateth and drinketh; thou canst not say that he is no loving god: therefore worship him.[6]

In the story, Daniel kills the dragon by poisoning it. But the *sirrush* and the problems it posed could not be gotten rid of so easily, for there it was, boldly emblazoned on the enameled bricks of the Ishtar Gate which Koldewey himself had unearthed.

Das kann man von den phantastischer gestalteten Mischbildungen der babylonisch-assyrischen Kunst nicht sagen.

4 Koldewey, *Das Ischtar-Tor,* p. 29.

5 Ibid., pp. 29.

6 *Bel and the Dragon,* pp. 23–24, Authorized King James Version.

The renowned naturalist Ivan T. Sanderson summed up Koldewey's conundrum this way:

> (Despite) his solid Teutonic background, Professor Koldewey became more and more convinced that it was not a representation of a mythical creature but an attempt to depict a real animal, an example of which had actually been kept alive in Babylon in very early days by the priests. After much searching in the depths of his scientific soul, he even made so bold as to state in print that this animal was one of the plant-eating, bird-footed dinosaurs, many types of which had by that time been reconstructed from fossil remains. He further pointed out that such remains were not to be found anywhere in or near Mesopotamia and that the "Sirrush" could not be a Babylonian attempt to reconstruct the animal from fossils. Its characters, as shown in Babylonian art from the earliest times, had not changed, and they displayed great detail in scales, horns, wrinkles, the crest and the serpentine tongue, which, taken together, could not all have been just thought up after viewing a fossilized skeleton.[7]

So there it was, and the conundrum was extraordinary, any way one sliced it.

Lest it have been missed, however, it behooves us to retrace the steps of Koldewey's logic in order to exhibit the conundrum with the full force of its implications:

1) There were no fossil remains near Babylon by which the Babylonians could have artistically reconstructed such a fantastic creature;
2) The closest dinosaur resembling the *sirrush* was the iguanodon, but again, there were no remains of such a creature near Babylon that would have allowed an artistic reconstruction;
3) The *sirrush* appeared throughout Mesopotamian art with amazing consistency, whereas other mythological and chimerical creatures depicted in the art of the region varied over time;
4) The *sirrush* appeared in a context with other really existing animals, namely, the now-extinct aurochs; and thus,
5) Either the Babylonians managed to encounter some sort of dinosaur long after they were supposed to be extinct; *or,*
6) The *sirrush*, notwithstanding a generalized resemblance to the

7 Ivan T. Sanderson, *More "Things"* (Kempton, Illinois: Adventures Unlimited Press, 2007), p. 22.

> iguanodon, was some *other* sort of bizarre and chimerical creature unknown to modern paleontology, but nevertheless, really existing.

And to top it all off, the creature may have even been the basis for a famous story from the biblical Apocrypha.

Robert Koldewey, 1855–1925

Left View of the Sirrush

However, while Professor Koldewey was busily digging up all sorts of problems for standard academic fundamentalisms of ancient history and the evolution of life, yet another German was posing problems of a different sort, for a very different sort of fundamentalism.

B. Delitzsch's Dilemma: Babel und Bibel

Friedrich Delitzsch (1850–1922) was a noted German Assyriologist who had the distinction of having caused an international firestorm of controversy that it took no less than the efforts of Kaiser Wilhelm II, acting in his capacity as the chief bishop of the German Lutheran Church, to stamp out.

The controversy began innocently enough.

The Cambridge scholar C.H.W. Johns, in his 1903 "Introduction" to Delitzsch's lectures, summarized its rather innocent beginnings in the following way:

> The announcement that Professor Friedrich Delitzsch, the great Assyriologist, had been granted leave to deliver a lecture upon the relations between the Bible and the recent results of cuneiform research, in the august presence of the Kaiser and the Court, naturally caused a great sensation; in Germany first, and, as a wider circle, wherever men feel interest in the progress of Science. The lecture was duly delivered on the 13th of January 1902, and repeated on the 1st of February.
>
> Some reports of the general tenour of the discourse reached the outside world, and it was evident that matters of the greatest interest were involved. In due course appeared a small book with the text of the lecture, adorned with a number of striking pictures of the ancient monuments. This was the now celebrated *Babel und Bibel.* [8]
>
> The title was a neat one, emphasizing the close relation between the results of cuneiform studies and the more familiar facts of the Bible.[9]

One may easily imagine the scene: the Kaiser resplendent in his uniform, his marshals and ministers surrounding him, sitting in ornamented chairs, listening to the distinguished professor elaborating his discoveries and conclusions.

But then, according to Johns, events took a decidedly strange turn. Indeed, "it came, therefore, as a shock of surprise to find that rejoinders were being issued." That wasn't all:

---

8 i.e., *Babel and Bible* (ed.).

9 C.H.W. Johns, "Introduction," to Friedrich Delitzsch, *Babel and Bible* (Eugene, Oregon: Wipf and Stock, 2007), pp. v–vi.

> A rapid succession of articles, reviews, and replies appeared in newspapers and magazines, and a whole crowd of pamphlets and books. These regarded the lecture from many varied points of view, *mostly with disapproval.* The champions of the older learnings assailed it from all sides. Even those who had been forward to admit nothing but a human side to the history and literature of Israel were eager to fall on the new pretender to public favour; and, to the astonishment of many, these arose a literature *zum Streit um Babel und Bibel.*
>
> As the echoes of this conflict reached our ears, we seemed to gather that the higher critics, usually known for their destructive habits, were now engaged in defending, in some way, the Bible against the attacks of an archaeologist and cuneiform scholar. This seemed a reversal of the order of nature. We had been used to regard the archaeologist, especially the Assyriologist, as one who had rescued much of the Bible history from the scepticism of literary critics.[10]

But then, to make matters even worse, Delitzsch was invited to deliver yet another lecture in the presence of Kaiser Wilhelm and his court.

Kaiser Wilhelm II von Hohenzollern

10 C.H.W. Johns, "Introduction," to Friedrich Delitzsch, *Babel and Bible*, pp. vii–viii, emphasis added.

And that was when the "cuneiform hell" broke loose, requiring the Kaiser's personal imperial intervention to quiet the controversy:

> But now reports of a very disquieting nature reached us. Our papers had it from their correspondents that a very direct attack was made on Holy Scripture, and even, it was not obscurely hinted, on the fundamental doctrines of the Catholic Faith. The storm broke out afresh in Germany, and spread hither also. We learnt, to our amazement, not exactly realizing the Kaiser's position as *Summus Episcopus*, that he had seen fit to address a letter, the text of which appeared in the *Times* of February 25th.
>
> That lectures, even on such an interesting subject, could lead to measures of such high state policy was a guarantee that the matter had passed beyond the circles of scholarship and research, and was become a matter of national concern. We could not afford to remain longer in ignorance of what had stirred our allies so profoundly.[11]

Just exactly *what* had Delitzsch said to the Kaiser and his court, and why did it have everyone from "bible believer" to "higher critical skeptic" so exercised that it required a letter from the Kaiser himself, in his capacity as "highest bishop" in the Lutheran Church, to quell?

If one glances at the short biographical sketch of Delitzsch in the online encyclopedia Wikipedia, one begins to have some approximation:

> Friedrich Delitzsch specialized in the study of ancient Middle Eastern languages, and published numerous works on Assyrian language, history and culture. He is remembered today for his scholarly critique of the Biblical Old Testament. In a 1902 controversial lecture titled "Babel and Bible," Delitzsch maintained that many Old Testament writings were borrowed from ancient Babylonian tales, including the stories of Creation and the Great Flood. During the following years there were several translations and modified versions of the "Babel and Bible." In the early 1920s, Delitzsch published the two-part *Die große Täuschung* (*The Great Deception*), which was a critical treatise on the book of Psalms, prophets of the Old Testament, the invasion of Canaan, etc. Delitzsch also stridently questioned the historical accuracy of the Hebrew Bible and placed great emphasis on its numerous examples of immorality....[12]

---

11 Ibid., pp. x–xi.

12 en.wikipedia.org/wiki/Friedrich_Delitzsch

Clearly, Delitzsch's program was a total one, as his work entitled *The Great Deception* implies, but that program was first enunciated in his lectures before the Kaiser and his court, the lectures which eventually became *Babel and Bible.*

Delitzsch himself put the matter this way at the very beginning of his first lecture:

> What is the object of these labours in distant, inhospitable, and dangerous lands? To what end this costly work of rummaging in mounds many thousand years old, of digging deep down into the earth in places where no gold or silver is to be found? *Why this rivalry among nations for the purpose of securing, each for itself, these desolate hills — and the more the better — in which to excavate?* And from what source, on the other hand, is derived the self-sacrificing interest, ever on the increase, that is shewn on both sides of the ocean, in the excavations in Babylonia and Assyria?
>
> To either question there is one answer, which, if not exhaustive, nevertheless to a great extent tells us the cause and aim: it is the Bible. [13]

Observe carefully both what Delitzsch has *implied*, and what he has *actually said* here.

First, Delitzsch has implied that the excavations in Mesopotamia had a direct bearing on our understanding of the origins of our Judeo-Christian civilization and, to a lesser extent, the civilization of the Islamic world as well. But secondly, and more importantly, he has explicitly stated that the *control* of such excavation sites was a matter of great power rivalry, since it was precisely those nations that were aiming to control such sites across the Middle East, and *that* implies that there might be a hidden agenda at work behind the seemingly innocent purposes of archaeological digging. Just what that agenda may be will become more and more evident in a moment.

Delitzsch interlarded the actual published version of his lectures with a number of pictures that clearly pointed to *some* deep Babylonian-Assyrian-Sumerian origin or influence upon much of the stories of the Old Testament:

> May we point to an old Babylonian cylinder-seal? Here, in the middle, is the tree with hanging fruit; on the right the man, to be recognized by the horns, the symbol of strength, on the left the woman; both reaching out their hands to the fruit, and behind the

13 Friedrich Delitzsch, *Babel and Bible*, pp. 2–3, boldface emphasis Delitzsch's, italicized emphasis mine.

woman the serpent. Should there not be a connection between this old Babylonian representation and the Biblical story of the Fall?[14]

He then reproduces this depiction of the impression of the cylinder-seal:

Babylonian Cylinder-Seal Depiction of the Fall of Man[15]

There were other suggestive artistic parallels between popular Christian imagery and ancient Babylonian and Assyrian art. For example, there was a depiction of Assyrian king Ashurbanipal slaying a lion, all too eerily similar to depictions of St. George slaying the dragon:

King Ashurbanipal Slaying a Lion[16]

14 Friedrich Delitzsch, *Babel and Bible*, p. 56.

15 Ibid.

16 Ibid., p. 22.

The theme was repeated, in a different guise, with the god Ninurta/Marduk wielding powerful thunderbolts to slay a chimerical "dragon":

Ninurta/Marduk Slaying the Dragon with Thunderbolts[17]

Delitzsch's commentary is worth citing:

> It is interesting to note that there is still an echo of this contest between Marduk and Tiamat in the Apocalypse of John, where we read of a conflict between the Archangel Michael and the "Beast of the Abyss, the Old Serpent, which is the Devil and Satan." The whole conception, also present in the story of the knight St. George and his conflict with the dragon, a story brought back by the Crusaders, is manifestly Babylonian.[18]

As I have argued elsewhere, the struggle between Marduk and Tiamat might actually be the dimly recollected memories of an actual cosmic, or interplanetary, war fought in very ancient times with extraordinary technologies.[19] In Delitzsch's time, with the discoveries of Nikola Tesla and other inventors already transforming the world, it would have been an easy step for the elites of those great powers to read the ancient Sumerian, Babylonian, and Assyrian texts, and to conclude that they contained hints of a lost technology of vast power, and hence, one has one explanation for the possible hidden motivations of the scramble those nations showed to control the various archaeological sites.[20]

---

17 Ibid., p. 52. Q.v. my *The Cosmic War* (Adventures Unlimited Press), pp. 53–58 for a physics rationalization of the peculiar shape of these "thunderbolts." Authorities appear to be divided as to which god — Ninurta or Marduk — is depicted in the relief.

18 Ibid., p. 52.

19 Q.v. my *Giza Death Star Destroyed* (Adventures Unlimited Press), pp. 37–52; and *The Cosmic War* (Adventures Unlimited Press).

20 The struggle appears to have continued to modern times with the recent episode of the looting of the Baghdad Museum. Q.v. my *Nazi International* (Adventures Unlimited Press), pp.

1. The Cuneiform Tablets and an Out-of-Place Name for God

But by far the most sensational piece of evidence and commentary that Delitzsch produced in his lecture was a set of cuneiform tablets.

> What is there to be seen on these tablets? I shall be asked. Fragile, broken clay upon which are scratched characters scarcely legible! That is true, no doubt, yet they are precious for this reason: they can be dated with certainly, they belong to the age of Hammurabi,[21] one in particular to the reign of his father Sin-mubalit. But they are still more precious for another reason: they contain three names which, from the point of view of the history of religion, are of the most far-reaching importance...[22]

...and here, he places a photograph of the tablets:

Delitzsch's Photograph of Three Cuneiform Tablets Dating from the time of King Hammurabi[23]

What was of interest to Delitzsch — and therefore to us — were three cuneiform names that were the heart of the controversy between the famous Assyriologist on the one hand, and all the skeptical higher critics and biblical fundamentalists on the other:

417–420, 232. See also my article "The Baghdad Museum Looting: Some More Thoughts" at www.gizadeathstar.com.

21 Hammurabi was the sixth king of Babylon, ca. 1792–1750 B.C.

22 Friedrich Delitzsch, *Babel and Bible*, p. 71.

23 Friedrich Delitzsch, *Babel and Bible*, p. 71.

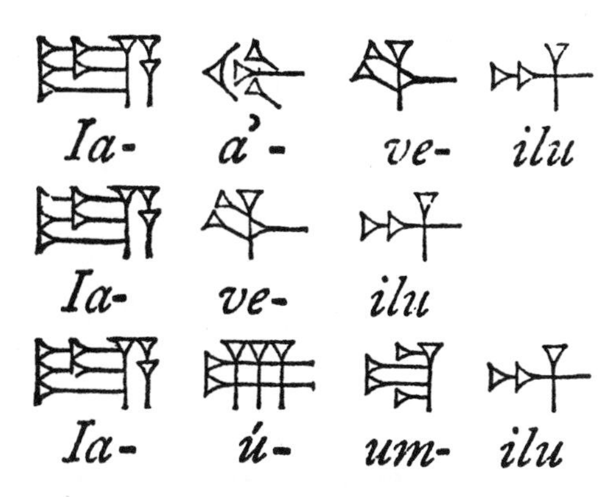

The Three Cuneiform Names at the Heart of the Controversy[24]

Notably, the three names contained the root word "Ia" and in one significant instance, the name "Iave" or, to Hebraicize it, "Yahveh," the very proper name of God, the mysterious "tetragrammaton" which was, according to the account in Exodus 3, spoken or revealed to Moses after the Exodus from Egypt some centuries later! In short, there was nothing special about the name "Yahweh," or "Iave."

2. The Documentary Hypothesis: Astruc to DeWette

One may get a sense of the enormous implications of Delitzsch's discovery — at least for the biblical literalists of the period — by posing an obvious question: what was a supposedly uniquely Hebrew proper name for God doing in cuneiform texts manifestly much older than the book of Exodus, and in a very *un*-Hebrew, very *Sumerian* context?

But what of the problems it posed to the literary higher critics? Why did *they* take umbrage at Delitzsch's cuneiform tablets? To answer that question requires a short excursion into a critical theory called "the Documentary Hypothesis," or as it is also sometimes known, the "Graf-Wellhausen Hypothesis" or the "JEDP theory." In its recognizable modern form, this theory holds that the first five books of the Old Testament — the "Pentateuch" or "Torah" — were composed from different underlying documentary "sources" indicated by the four letters J, E, D, and P. The theory began in Enlightenment France with the observations of the French physician Jean Astruc.

Astruc noticed that in the Hebrew text of the first two chapters of the Book of Genesis, each chapter referred to God by a different name, *Elohim* (אלהים), translated "God" in the Authorized version, was the name used in

24 Ibid.

chapter one, and *Yahweh* (והיה), translated "LORD" (in all capitals) in the Authorized version in chapter two. In order to account for this difference, Astruc reasoned that Moses, when composing the "creation accounts" (which he assumed both chapters represented), had in fact utilized two independent sources, or "documents."[25] In so arguing, he provided the metaphysical and philological first principle that would guide subsequent scholarship to elaborate the fully fledged Documentary Hypothesis: *different Divine Names indicate the presence in the extant text of different underlying source material for that text.*

By 1853, nearly a century later, the German critic Herman Hupfeld would extend this principle to its logical conclusion: *differences within passages of overall style or vocabulary constituted a sufficient basis upon which to posit different underlying documentary sources from which those stylistic differences derive.* With "the Astruc Principle" and the "Hupfeld Corollary," a critical agenda of its own was emplaced and empowered, for now the various names of God could come, with a certain brazen and nominalistic elegance, to stand for something completely mundane rather than for some characteristic metaphysical property of God; they came, within the historical phenomenology of the hypothesis itself, to stand only for the source documents from which the final extant text was alleged to derive. The divine names, so to speak, were only the revelations of no-longer-extant source documents, which were the task of critical scholarship to discern and disentangle. And the Germans, more than anyone else, were the ones most busily engaged in this process.

It is worth pausing to consider the implications of all of this as possible manifestations of yet another agenda. By empowering the critic himself, with all his specialized tools of knowledge of the original languages, philology, and other ancient texts, a complete end run was done around existing ecclesiastical magisteria and doctrines, and additionally, the entirety of the Old Testament came to be viewed within such circles as the special creation over centuries of the Hebrew priesthood and elite, with the occasional bow to Egyptian origins for much of it.[26]

In any case, once the first two chapters of Genesis had been subjected to the "Astruc Principle" and the "Hupfeld Corollary," there was nothing logically to prevent their application to other passages of the Torah.[27] Indeed,

---

25 Gleason L. Archer, *A Survey of Old Testament Introduction* (Chicago: Moody Press, 1964), pp. 73–75; R.K. Harrison, *Introduction to the Old Testament* (Grand Rapids, Michigan: William B. Eerdmans Publishing Company), pp. 11–12.

26 Critics would point out, for example, that there was nothing *Hebrew* about the name "Moses," but rather, that the name was properly Egyptian in origin, as in the name of the pharaoh Thutmose. Other critics would point to the close similarity of the biblical decalogue to similar statements in the Egyptian *Book of the Dead.*

27 Archer, *A Survey of Old Testament Introduction,* pp. 77–78; R.K. Harrison, *Introduction to the Old Testament,* p. 17.

it was Johann Gottfried Eichhorn who first extended Astruc's criterion of the divine names as indicating separate source documents to the remainder of the book of Genesis and on into the first two chapters of the book of Exodus, in his *Introduction to the Old Testament,* published in Germany between 1780 and 1783. This work earned him his lasting epithet as being the "father of Old Testament criticism."[28] What was new with Eichhorn was the coupling of Astruc's philological principle with the new assumption that Moses had not authored any of the Torah or "Five Books of Moses." In other words, it was Eichhorn who in fact accomplished the empowerment of the critical scholar and the accompanying agenda, for if Moses did not author those books, and they were, on the contrary, the editorial compilation from sources made over time, then it followed that a massive task of historical reinterpretation and reconstruction would have to be undertaken. In Eichhorn's case, the "ancient agenda" at work in the text was simple: he maintained that the ancient Hebrew theology had evolved or developed from a primitive polytheism to an advanced personal monotheism, an evolution that in turn implied a post-Mosaic date for the emergence of the Torah in its final textual form.

Once the Torah was no longer the work of "Moses," or, to put it differently, one author, the way was then clear for critics to question the compositional, and therefore, the metaphysical and moral unity and integrity of the Torah. Indeed, as the elaboration of the Documentary Hypothesis proceeded throughout the late eighteenth century and all throughout the nineteenth, as the presupposition of unitary authorship collapsed, the discovery of textual, moral, and metaphysical contradictions within it grew in inverse proportion. With Eichhorn, then, we have Astruc's division of two "sources," the J or Jahwist source, and the E or Elohist source, extended to the entirety of the book of Genesis and on into Exodus chapters one and two.

One of the first to pursue the implications of Eichhorn's abandonment of Mosaic authorship was Wilhelm M.L. DeWette in the first half of the nineteenth century. He maintained that the Book of the Law which was discovered in 621 B.C. during King Josiah's reign, as recounted in II Kings 22, was in fact the book of Deuteronomy. DeWette argued that, since King Josiah and the high priest Hilkaiah were concerned to abolish localized sanctuaries and places of sacrifice and to centralize worship in the Jerusalem Temple, then, so his argument ran, the book which was "discovered" had in fact been deliberately composed for that purpose by an agent of the Temple, and its discovery was staged at the appropriate moment. For DeWette, the whole production, in other words, was in the service of a hidden agenda, namely,

28 R.K. Harrison, *Introduction to the Old Testament,* p. 14.

to centralize worship, thus solidifying the kingdom, and enriching the royal and temple treasuries. Readers of my previous book, *Babylon's Banksters,* will recall that it was precisely in alliance with ancient temples that the ancient banking fraternity often hid its own agendas.[29] But as we shall see in chapter three, there are possibly even more hidden, technological agendas at work in this maneuver. In any case, this pinpointed the date of the book of Deuteronomy to 621 B.C.[30] With DeWette's "discovery," the third document, D for the "deuteronomist" document, had been found. We now have J, E, and D.

3. The Documentary Hypothesis: Hupfeld's "Copernican Revolution"

With the work *Die Quellen der Genesis (The Sources of Genesis)* by the aforementioned Herman Hupfeld in 1853, the "Copernican revolution" in the history of the Documentary Hypothesis occurred.[31] His contribution to the evolution of the hypothesis were three new principles of examination of the Torah, and a newly discovered source document, so it is worth taking some time to examine Hupfeld in closer detail.

The first component in Hupfeld's "Copernican revolution" was that, by subjecting the previously isolated "E" document to a new philological examination, he discovered there were portions of "E" which, with the exception of the divine name "Elohim" itself, otherwise greatly resembled the "J" document in diction, style, and thematic focus. Thus, there were some portions of "E" which, in the historical scheme of the emerging hypothesis, appeared to be material as early as that of "J." Hupfeld was obliged, therefore, to separate "E" into *two further documents,* an earlier "E" that resembles "J," which he called "E_1," and the rest of "E" which became "E_2," or more simply, the original "E." There were now four documents, displaying more or less the following chronological order: E_1EJD. It was this E_1 that became the later "priestly" or "P" document, and it will henceforth be referred to as such. With this change in symbols, the order of documents now reads PEJD.

As if this were not already confusing enough, to this apparatus Hupfeld added a second, and perhaps the most important, principle in the arsenal of presuppositions of Old Testament criticism, for he maintained that the documents thus distinguished by the criteria of different vocabulary, diction, and interest had *integrity;* that is, they could not only be *distinguished* within the extant text of the Torah, but actually *recovered and reconstructed* as separate

29 See my *Babylon's Banksters* (Port Townsend, Washington: Feral House, 2010), pp. pp. 159-204.

30 R.K. Harrison, *Introduction to the Old Testament,* p. 15. In De Wette's case, other factors are at work, including a reaction against critical rationalism, q.v. John Rogerson, *Old Testament Criticism in the Nineteenth Century* (London: S.P.C.K., 1984), pp. 28–49.

31 Gleason L. Archer, *A Survey of Old Testament Introduction,* p. 77.

documents from that text. The text, in short, could be *rewritten.* (Talk about agendas!) This assumption was posited no doubt out of the perceived need on the part of Old Testament criticism to justify its increasingly radical reconstructions of early Hebrew history with a measure of "scientific" verifiability.

Thirdly, Hupfeld posited the existence of an "editor," or *Redaktor*, designated as "R" in the growing non-propositional calculus of the critic, an "R" who edited the portions of the text in "E" which resembled "J" which, according to the theory, should have belonged in "J" to begin with, except for the fact of the presence of the name *Elohim* and not *Yahweh!* If this is not confusing enough, it can be restated to make the confusion even more explicit: "R" lay behind the conflation of "P" and "E_2" into "E." This "R" was therefore truly a godsend, a literal *Redaktor ex machina*, for "whenever the theory ran into trouble with the facts or ran counter to the actual data of the text itself, the bungling had of R (the anonymous redactor) was brought in to save the situation."[32] This assumption of the redactor indicates the dialectical impasse into which the theory was quickly coming, for the discovery of "P" contradicted Hupfeld's own assumption of the integrity and recoverability of the source documents; for if "P" which so closely resembled "J" in diction could only be posited by positing a redactor — who is introduced precisely in order to account for the *resemblances* between the "documents" — then the integrity of the source documents collapses, for that integrity is dependent upon the *differences* in style and vocabulary to begin with! In other words, Hupfeld's redactor fills precisely the same function as an *author* utilizing sources, which is where the theory began.

4. The Documentary Hypothesis: Karl Heinrich Graf and Julius Wellhausen

The ultimate *reductio* was reached in the work of Karl Heinrich Graf in 1866. As Hupfeld had divided "E" into "P" and "E," Graf in turn distinguished within P between two *further* sources. He claimed to have discovered material in "P" which could only have been written *after* the deuteronomic legislation allegedly discovered by King Josiah in 621 B.C. Thus, there was some "P" that could only have come from a period later than "D." However, there was also historical material in "P" that could only have come from a period *earlier* than "D." Thus, P (Hupfeld's E_1) became "P_1" and "P_2." Thus, the chronological order from earliest to latest layers of source documents was P_1 (the really old P), E, J, D, and P_2 (the legalistic or priestly P).

Graf's separation of "P" into two different sources of "P" was not to endure. In 1869 the Dutch critic Abraham Keunen argued that P was a unified

32 Gleason L. Archer, *A Survey of Old Testament Introduction*, p. 78.

source, because the legal and legislative portions could not be separated from the historical ones without resulting in complete confusion, since the proper understanding of the one required the other. But, since Graf had fairly well established that the priestly legislation occurred after the Babylonian Exile, the entire P document had to have been composed after the discovery of "D" in 621. *This was a "gentle critical way" of saying that most of the laws in the Torah were really the product of a post-Exilic priesthood.* It was a gentle way of saying that, rather than those laws being the basis for Talmudic Judaism, the reverse was true; Talmudic Judaism was the basis of those Torah laws. It was a gentle way of stating that the Torah had been composed as the result of an *agenda* by an *elite*, an elite that was, in its turn, continuous within Hebrew society throughout several centuries.

In any case, with Graf's overturning of the separation of P and its redating to a period after D, only one final component needed to be added, and that was the component that the J document was the product of the Hebrews when they worshipped their local tribal god, Yahweh, and thus, the order of documents came into its now standardized chronological order: JEDP. It was this order that Julius Wellhausen gave in his massive reconstruction of ancient Hebrew history in his now classic work on the subject, *Introduction to the History of Israel.*

5. The Critical Suspicion of an Agenda

To put it succinctly, and regardless of its many problems, the higher critics who elaborated the Documentary Hypothesis were perhaps correct on one thing: they suspected that behind the biblical stories and the pious reasons advanced in them for various events, that there lay a hidden agenda of a hidden elite, manifesting itself, ever so faintly to be sure, in the philological and thematic differences between various passages. For these critics, many of whom had ceased to believe in any sort of God, criticism and the suspicion of "an agenda" at work in the texts became a way that they could plausibly read those texts and retain them in the canon of Western culture. If one accepts for the sake of argument that this is true, then perhaps they were too oblivious to the implications of their own insight, for such unity as the Torah had came not from an individual author nor even from the clumsy — yet "infallible" — hand of the *Redaktor ex machina*, but from the enduring influence and extent within ancient times of a surviving post-cosmic war elite.

6. The Suggestions of an Agenda at Work with the Critics: Weishaupt's Strange Comment and a Hidden Illuminist Role in Early Old Testament Criticism?

The critics might have looked closer to home for the possible work of an agenda in their own midst, in addition to one behind the texts they were scrutinizing, and that possible agenda may be pointed out by noting that higher criticism of the Old Testament, particularly in the hands of an Eichhorn, began in southern Germany in the eighteenth century, a time, and place, rife with what would become a by-word for conspiracies and agendas: the Illuminated Freemasonry of University of Ingolstadt professor of Canon Law, Adam Weishaupt, better known as the founder of the Bavarian Illuminati.

However, there do exist serious grounds for considering a possible relationship between the two. Both Masonry and criticism at that time shared a basic presupposition with respect to the Pentateuch, namely, that the ancient history of man and of Israel as recorded there was at best an allegorical compilation from other, earlier traditions, traditions either deliberately omitted or obscured by the extant biblical text.

More importantly, it was the stated goal of Adam Weishaupt and his Illuminati "to acquire the direction of education — of church management — of the professorial chair, and of the pulpit" and to "gain over the reviewers and journalists" in order to spread Illuminist opinions, and thus it would seem highly likely that biblical studies would have been a principal target for this agenda.

Even more importantly, Weishaupt himself refers once, in the writings captured by the Bavarian government, to his own "*history of the lives of the Patriarchs*" which, though apparently no longer extant, does indicate his own deep interest in biblical studies, and knowing Weishaupt's predilections, we may easily guess that this history was hardly in line either with Catholic or Lutheran orthodoxy of the period.

The Illuminati did indeed make great strides in the recruitment of prominent clergy for the day. Captured Illuminist writings list as members of the order one "Baader, professor," one "Barhdt, clergyman," and a "Danzer, canon," who most likely are Franz Xavier von Baader (1765–1841), Karl Friedrich Barhdt (1741–1792), professor of biblical archaeology at Erfurt, and Jacob Danzer (1743–1796), professor of moral and pastoral theology at Salzburg, a haven of Illuminati activities.[33]

But perhaps most importantly, Weishaupt boasts in one of his letters, "Who would have thought that a Professor at Ingolstadt was to become the teacher of the professor of Goettingen, and of the greatest men in Germany?"

33 John Robinson, A.M. *Proofs of a Conspiracy* (New York: George Forman, 1798), pp. 160–161.

Goettingen, at the time that Weishaupt wrote his letter, was home to Old Testament professor Michaelis, and his more celebrated pupil, Eichhorn.

7. The Explosive Thunderclap of Delitzsch's Dilemma

So, why *was* "Delitzsch's Distinctive Dilemma" such an explosive thunderclap that its echoes reverberate down to our own times in obscure scholarly debates?

So how then may one summarize "Delitzsch's Distinctive Dilemma"?

1) On the one hand, the clear implication of Delitzsch's cuneiform names containing the root "Ia" and even "Iave" centuries before the epiphany to Moses at the Burning Bush, as recounted in Exodus chapter 3, is that there was nothing inherently *special* about the name "Yahweh" itself, since it was known long before the Exodus, and that in turn demoted the Exodus passage from the status it always had had within within Jewish and Christian theology as a special monotheistic revelation, challenging conservative literalist fundamentalisms; and,
2) On the other hand, the presence of the name in tablets clearly datable to a time period from Hammurabi also meant that the careful chronological reconstructions of the Documentary Hypothesis — *the critics' 'new fundamentalism'* — were on very shaky foundations at best, or clearly dubious and spurious at worst.

But there was a further implication:

3) The presence of such names in very old cuneiform tablets also implied that the biblical text was indeed *edited*, but in a very different way, and for a very different purpose, than that proposed by the higher critics advocating the Documentary Hypothesis. Indeed, as Delitzsch himself pointed out, the presence of "Cosmic War" and "Fall of Man" themes in Babylonian art, themes paralleled in the biblical stories, suggested that the editing was wholesale and present throughout the region's texts and artwork. It suggested, in other words, that there was an agenda at work throughout *all* texts from the region — biblical or otherwise — and that to learn what that story and agenda was, one would have to reconstruct that history by a careful critical process.

When we put these considerations into the implications of Koldewey's musings on the *sirrush*, we get an even further expanded list of implications, for not only are biblical history and the wider history of religion and culture affected, so too, and by the same token, *is the history of science and technology itself.* This implies that the "agenda" referred to previously, the agenda at work in the careful editing of texts, may be trying to hide something about four things: God, man, religion, and science itself.

Koldewey's *sirrush* with its odd and bizarre mixture of serpentine, ornithological, and feline characteristics points the way, for with the modern science of genetics, the creation of such chimerical creatures looms as an ever more feasible reality, a reality that the ancient texts from Mesopotamia also implied once existed, and even implied the use and manipulation of a technology to achieve it. Could it all have been an agenda of massive misdirection, a case of sleight-of-hand designed to get the vast majority of mankind looking elsewhere, while an informed elite, looking at ancient texts and seeing a lost science and technology, was really digging and scratching in the desert sands in an attempt to recover and redeploy that lost technology?

That possibility informs the rest of this book, and to it, we now turn.

Two

Marduk Measured the Structure of the Deep:

Megalithic Measures and the Sumerian "Reform"

∴

"And [Marduk] measured the structure of the deep."
—*Enuma Elish*, Tablet 4[1]

"In reality there are no mainstream measuring systems in common use today that fail to owe a debt to the original, integrated megalithic system."
—Christopher Knight and Alan Butler[2]

IF ONE IS TO BELIEVE the ancient Babylonian "cosmic war" epic, the *Enuma Elish*,[3] then almost immediately upon the conclusion of that war, Marduk, the victor over the evil villain Tiamat, set out to "measure the structure of the deep." The mind's almost subconscious reaction to this bit of information is to chuckle and write the statement off as yet more proof that the *Enuma Elish* is nothing but a bit of ancient Mesopotamian science fiction imagination run amok, with no basis in the realities of the necessities of a postwar "cleanup" of the rubble.

1 *Enuma Elish*, Tablet 4, ed. L.W. King, F.S.A., Vol. 1 (London: Luzac and Co., 1902), p. 77.

2 Christopher Knight and Alan Butler, *Before the Pyramids: Cracking Archaeology's Greatest Mystery* (London: Watkins Publishing, 2009), p. 34.

3 See my *Giza Death Star Destroyed* (Kempton, Illinois: Adventures Unlimited Press, 2005), pp. 37–49, and my *The Cosmic War* (Kempton, Illinois: Adventures Unlimited Press, 2007), especially pp. 150–166. I strongly dissent from the standard academic view that this text represents an allegory of cosmology and creation.

Nothing, however, could be further from the truth.

Before proceeding to uncover that "truth," however, it is necessary to reiterate three of the assumptions being made in this book that bear directly on our examination of the subject matter of this chapter.

Firstly, we have assumed that an "elite" survived that war, scattered throughout various places on planet Earth, and possibly elsewhere in local space. Secondly, we have assumed that those elites had agendas, both open and hidden, and that among the open ones were the quickest possible re-establishment of the basic necessities of civilization and the quickest possible global extension of them. Given the devastation caused by that ancient "cosmic war," however, this goal necessarily had to operate over centuries and millennia of painstaking effort. Such an effort and long-range goal thus required that these surviving elites had to organize themselves — *and their knowledge* — in such a fashion to be able to preserve that goal and the knowledge base *even if the foundations of that knowledge were lost in the short term.* This in turn required what I have referred to as the creation of a "unified intention of symbol," by which myths and stories were created to encode and preserve that science, technology, and history in multi-layered, complex imagery and symbols that would be decodable once mankind had reached a similar pitch of scientific and social development as had been the case in the prewar civilization.[4] The possible *hidden* agendas have already been alluded to in the previous chapter, for "Koldewey's Conundrum" and "Delitzsch's Dilemma" point to the possible hidden manipulations of surviving technologies, and very possibly of religion itself, as some of the means to accomplish this long-term goal.

Thus we arrive at the third and final component of the assumptions that bear directly on this chapter, and that is that some fragments of the technology and scientific knowledge, howsoever rudimentary, survived that ancient cosmic war and were put to immediate use to re-establish and preserve those elements of civilization necessary for human survival and progress.

This assumption highlights in stark relief the dangerous "gap of history" in the standard academic models of human prehistory, for as many people know, the remarkably stable, advanced, and highly integrated societies of Egypt and Sumer sprang up rather suddenly and with very little antecedent preparation. They were just suddenly *there.* It was into this "gap of history" that an Oxford professor stepped, showing that there had indeed been "preparation," and that this preparation had been done by an elite, that it was deliberately coordinated, and remarkably consistent in its spread over Europe and the Middle East.

4 See my *The Giza Death Star Destroyed,* pp. 48–52, and *The Cosmic War,* pp. 75–80.

A. An Oxford Professor Overturns the Standard Model: Alexander Thom, His Work, and Its Implications

That professor's name was Alexander Thom, and the work he did was nothing less than brilliant. So brilliant was it, in fact, that his results continue to be questioned (a polite euphemism for "rejected") by the history, anthropology, and archaeology communities within academia. Indeed, Thom was not a professional historian, anthropologist or archaeologist, and that was part of the problem. He was an *engineer* and a mathematician, and like engineer Christopher Dunn who would so vex Egyptologists and anthropologists by maintaining that the Great Pyramid was a machine,[5] Thom's conclusions would fly wholesale in the face of the revered assumptions of anthropology, archaeology, and historiography regarding the ancient pre-classical history of mankind. Thom's work is relatively difficult to come by, but British authors Christopher Knight and Alan Butler are two researchers who have studied it thoroughly and popularized his findings, expanding on them and adding considerable insights of their own, and accordingly their works will be followed closely here.

Born in the year 1894 in Scotland, Thom attended the University of Glasgow and later returned there as an engineering lecturer. During World War II Thom did work for the British government, and after the war was over, returned to the academy, this time assuming a post as a Professor of Engineering at the University of Oxford where he remained until he retired in 1961. Like so many others associated with the University of Oxford Thom developed an interest in the "pre-history" of man, suspecting that all was not well with the "standard model":

> Thom's interest in Megalithic structures began in his native Scotland, where he noticed that such sites appeared to have lunar alignments. In the early 1930s he decided to study some of the sites and began a process of careful surveying that was to take him almost five decades. In addition to his lecturing, Alexander Thom was a highly talented engineer in his own right and he taught himself surveying, which enabled him to look at more Megalithic sites — and in greater detail — than anyone before or since.[6]

In this concentration on close engineering and surveying observations of ancient structures, Thom was very much like another famous contemporary, Sir William

5 Christopher Dunn, *The Giza Powerplant* (Bear and Co., 1998).

6 Christopher Knight and Alan Butler, *Civilization One: The World is Not as You Thought it Was* (London: Watkins Publishing, 2004), p. 15.

Flinders Petrie, the surveyor par excellence of the Giza plateau in Egypt.

Thom's excavations and surveying work took him five decades, surveying sights

> from the gaunt and impressive standing-stone circles of the island of Orkney in the far north of Scotland, right down to the giant avenues of stones in their frozen march across the fields of Brittany in France... [where] Thom spent each and every summer for almost five decades carefully measuring and making notes. Together with family and friends and associates, he gradually built up a greater database, regarding megalithic achievement in building, than anyone before or since.[7]

Thom eventually published the results of his studies and surveying in an article written in 1951 for the *Journal of the British Astronomical Association.* The article was entitled "The Solar Observations of Megalithic Man." And over the next three years Thom expanded on this by adding three more articles for the *Journal of the Royal Statistical Society*, as well as publishing three books on his findings.[8]

Almost as soon as he had published his first article, the grumblings of discontent were heard from archaeologists and anthropologists, and by the time of his final articles and books, that grumbling had grown into a deafening silence after an initial flurry of stubborn rejection. As Knight and Butler state the case, "The difference in approach between Thom and the general archaeological community is fundamental,"[9] and boils down to a "collision of techniques."[10]

Well might archaeologists have complained, for in essence, what Thom did was to prove that the supposedly "primitive" megalithic builders of Britain were profound astronomers, surveyors, and engineers in their own right. Just precisely what had Thom done to earn the ire of archaeologists? He had made "a startling claim":

> He maintained that he had found that the structures left by late Stone Age man had been built using *a standard unit of measure that was so precise that he could identify its central value to an accuracy that was less than the width of a human hair.* The idea that these simple people

7 Christopher Knight and Alan Butler, *Before the Pyramids: Cracking Archaeology's Greatest Mystery* (London: Watkins Publishing, 2009), p. 28.

8 Knight and Butler, *Civilization One*, p. 16.

9 Knight and Butler, *Civilization One*, p. 20.

10 Ibid., p. 21.

> from prehistory could have achieved such accuracy flew in the face of all the worldview of most archaeologists.[11]

But that was not the *only* problem, for this unit of measure, spread throughout Brittany in France and the British isles, was consistent in that accuracy, and moreover, was based — as we shall see in a moment — on extremely accurate astronomical and geodetic observation. How could it be, then, that "the supposedly unsophisticated people of Stone Age Britain possessed a fully integrated system of measurement based on a deep understanding of the solar system"?[12]

This, of course, was the *real* problem, for it implied a sophisticated knowledge base from which the ancient sites were constructed, and that of course raised the dreaded possibility that something much more sophisticated by way of a civilization had pre-existed the Stone Age peoples. In short, Thom's work raised the dreaded possibility of the "A" word, of "Atlantis." In any case, this standard unit of astronomical and geodetic measure Thom called "The Megalithic Yard," and it became a banner around which supporters of the theory of a pre-existent and sophisticated civilization rallied. Indeed, its mere existence posed a kind of historiographical problematic for the standard views of human civilization's development, for "if he was wrong, the subject of statistics needs a fundamental reappraisal; but if his findings were reliable, the subject of archaeology needs equally careful reassessment. Further, if Thom was right, the development of human civilization may have to be rewritten!"[13] Knight and Butler quickly determined that the "only hope of resolving the issue, once and for all, was to attempt to find a reason why this length of unit would have had meaning for Neolithic builders, and then to identify a methodology for reproducing such a length at different locations."[14]

And here is where the mystery becomes even *more* intriguing!

1. Thom's Megalithic Yard and the Expansions of Knight and Butler

To see just how intriguing the mystery is, we must first know exactly what Thom's "megalithic measures" *were*.

> Thom identified the use of a standard unit he called a 'Megalithic Yard' (MY), which he specified as being equal to 2.722 ft +/- 0.002 ft (0.82966 m +/- 0.061 m). He claimed that there were also other related units used repeatedly, including half and double Megalithic

11 Ibid., p. 1, emphasis added.

12 Ibid., p. 1.

13 Knight and Butler, *Civilization One,* p. 25.

14 Ibid.

> Yards and a 2.5 MY length he dubbed a Megalithic Rod (MR). On a smaller scale he found that the megalithic builders had used a fortieth part of a Megalithic Yard, which he called a 'Megalithic Inch' (MI) because it was 0.8166 of a modern inch (2.074 cm). The system worked like this:
>
> 1 MI = 2.074 cm
> 20 MI = ½ MY
> 40 MI = 1 MY
> 100 MI = 1 MR.[15]

But there was not only system in this ancient structure of measure, there was amazing consistency over a very large area:

> The lifetime work of Alexander Thom and his rediscovery of the Megalithic Yard resulted in a stunning conclusion that created an immediate paradox — *how could an otherwise primitive people build with such fine accuracy? Why did they do it and how did they do it?* Thom made no attempt to answer these questions. He reported on his engineering analysis and left the anthropological aspects for others to explain. *He did comment that he could not understand how these builders transmitted the Megalithic Yards so perfectly over tens of thousands of square miles and across several millennia and he acknowledged that wooden measuring sticks could not have produced the unerring level of consistency he had found.*[16]

How indeed had they done it? And much more importantly, *why?*

The problem posed by these questions only deepens when one considers the standard archaeological and anthropological view of where ancient measures came from:

> According to Thom, the units he discovered were extraordinary because they were *scientifically exact.* Virtually all known units of measurement from the Sumerians and Ancient Egyptians through to the Middle Ages *are believed to have been based on average body parts such as fingers, hands, feet and arms, and were therefore quite approximate.*[17]

---

15 Knight and Butler, *Before the Pyramids,* pp. 28–29.

16 Ibid., p. 29, emphasis added.

17 Ibid., p. 17, emphasis added.

In other words, Thom's measurements, recorded over several thousand square miles in Britain and France, and recording an accuracy of this ancient unit of measure to within the width of a human hair, gave lie to the archaeological assumptions that ancient measures were based upon ever-varying body parts! *Something else entirely* was in play!

2. The Methods

a. Thom's "Ancient Bureau of Standards" Theory

This fact posed an intriguing problem, namely, could one "reverse engineer" the *method* by which supposedly primitive Megalithic builders had derived such an accurate and consistent measure?

One theory that attempts to do this is — for want of a better expression — the "ancient Bureau of Measures" theory. Thom himself was at a loss to explain the consistent accuracy of the Megalithic Yard over such a large theory, and at first proposed the "ancient Bureau of Measures" theory:

> This unit was in use from one end of Britain to the other. It is not possible to detect by statistical examination any differences between the values determined in the English and Scottish circles. *There must have been a headquarters from which standard rods* (a rod could be of two types, but in this context there are pieces of wood cut to represent the Megalithic Yard) *were sent out...* The length of rods in Scotland cannot have differed from that in England by more than 0.03 inch (*0.762mm*) or the difference would have shown up. If each small community had obtained the length by copying the rod from its neighbor to the south the accumulated error would have been much greater than this.[18]

But there was a major problem with this theory, and Knight and Butler are quick to point it out:

> At that time Thom's data could not be explained by any mechanism known to be available to the people of the late Stone Age other than to assume that all rods were made at the same place and delivered by hand to each and every community across Scotland and England. Eventually, he would find the unit in use from the Hebrides to

18 Alexander Thom, *Megalithic Sites in Britain* (Oxford: Oxford University Press, 1968), cited in Knight and Butler, *Civilization One,* p. 19.

> western France, which makes the central ruler factory theory look most unlikely. *He also found it impossible to imagine why these early communities wanted to work to an exact standard unit.*[19]

In other words, the unit was spread over too large an area for the "ancient Bureau of Standards" theory to account for it. We will leave the full exposition of the answer to the question of why such primitive Stone Age people would have wanted "to work to an exact standard unit" to a later point in this chapter.

Thus, effectively, we are back to square one: how *did* these "primitive" Stone Age peoples come up with such a unit of measure, and reproduce it with unerring accuracy over such a wide area? The answer to that, according to Knight and Butler, is rather astounding, and points in turn to a hidden and hardly "primitive" elite acting as a guiding hand, and working behind the scenes.

3. Celestial Geometries and the Pendulum Method

The question of why such a primitive people would have need for such an accurate unit of measure spread over such a wide area is, however, crucial for understanding *how* they reproduced it. Knight and Butler state this methodological problem in the following fashion: "We saw that the only hope of resolving the issue, once and for all, was to attempt to find a reason why this length of unit would have had meaning for Neolithic builders, *and then to identify a methodology for reproducing such a length at different locations.*"[20] Optimally, this meant that "what our Megalithic mathematicians needed was a method of reproducing the Megalithic Yard that was simple to use, very accurate and available to people dispersed over a large distance and across a huge span of time."[21] It was the classic engineer's optimalization problem, for whatever this method was, it had to be a method that also "ensured that the unit of length would not change across time or physical distance,"[22] and this meant of course that in all likelihood, the unit was founded on something with a fairly constant base "in the natural world"[23] that would not change over time or physical location.

---

19 Knight and Butler, *Civilization One,* p. 19, emphasis added. For additional commentary on the "transmission problem" of the megalithic measures, see Knight and Butler, *Before the Pyramids,* p. 29.

20 Knight and Butler, *Civilization One,* p. 25.

21 Knight and Butler, *Civilization One,* p. 32.

22 Ibid., p. 33.

23 Ibid.

That, of course, implied that the answer lay in the stars, and in the Earth, themselves. And if this be the case, then the most obvious unit immediately known to such "primitive" observers would be a "day," and this is where the ultimate basis of the method of reproducing an accurate unit of measure begins:

> There are various ways of defining a day and the two principal types are what we now call a 'solar' day and a 'sidereal' day. A solar day is that measured from the zenith (the highest point) of the Sun on two consecutive days. The average time of the Sun's daily passage across the year is called a 'mean solar day' — it is this type of day that we use for our timekeeping today. A sidereal day is the time it takes for one revolution of the planet, measured by observing a star returning to the same point in the heavens on two consecutive nights. This is a real revolution because it is unaffected by the secondary motion of the Earth's orbit around the Sun. This sidereal day, or rotation period, is 236 seconds shorter than a mean solar day, and over the year these lost seconds add up to exactly one extra day, giving a year of just over 366 sidereal days in terms of the Earth's rotation about its axis.
>
> *In short, anyone who gauged the turning of the Earth by watching the stars would know full well that the planet turns a little over 366 times in a year, so it follows that this number would have great significance for such star watchers. If they considered each complete turn of the Earth to be one degree of the great circle of heaven, within which the Sun, Moon and planets move, then they would also logically accept that there are 366 degrees in a circle.*
>
> There really are 366 degrees in the most important circle of them all — the Earth's yearly orbit of the Sun. *Anything else is an arbitrary convention. It seemed to us that this was so logical that the 360-degree circle may have been a later adjustment to make arithmetic easier, as it is divisible by far more numbers than the 'real' number of degrees in a year. In other words, the circle of geometry has become somehow detached from the circle of heaven.*[24]

Note carefully the implication of these remarks, for the *natural* system of a celestial and geodetic measure would involve some system of a circle of 366 degrees, while at some *later* point — largely for the sake of simplified arithmetical calculation — a modified or *tempered* system was put into place by "someone." This is a significant point and it will be taken up again later in this chapter.

24 Knight and Butler, *Civilization One,* pp. 27–28, emphasis added.

But how was this "original" 366-degree system derived by such primitive peoples? Seeking to reconstruct the thought processes of the Megalithic builders, Knight and Butler came to some very practical conclusions:

> [It] is highly likely that they would also realize that sunrises across the year move exactly like a pendulum. At the spring equinox (currently around 21 March) the Sun will rise due east and then rise a little further north each day until the summer solstice (21 June) at which point it stops and reverses its direction back to the autumn equinox and on to the winter solstice, by which time it will rise well into the south. The Sun's behaviour across a year, as viewed from the Earth, creates exactly the same frequency model as a pendulum. It displays a faster rate of change in the centre and slows gradually to the solstice extremes, where it stops and reverses direction.[25]

So the first problem was "to puzzle over the issue of how any unit of time could possibly be converted into a linear unit."[26] The answer lay in the motion of the Sun during a year: the pendulum. In a certain way, the same of course could be said of the motion of nearby planets, such as Venus, for by using the very primitive "machine" of a pendulum and choosing a fixed reference point in the heavens, and counting the beats or swings of the pendulum as a chosen star moved between fixed observation points on the horizon.

The pendulum was a ready-made, simple machine, easily within the technological capacities of Megalithic builders, and moreover, so closely tied to the "invariable" properties of the Earth itself that it formed the perfect basis for the accuracy of Megalithic measures over so wide an area. The reason for this is simple, for the pendulum directly links the gravitational field of the Earth, the notion of time as a beat frequency, and the conversion *of both to a linear measure:*

> The time it takes a pendulum to swing is governed by just two factors: the mass of the Earth and the length of the pendulum from the fulcrum... to the centre of gravity of the weight. Nothing else is of significant importance. The amount of effort that the person holding the pendulum puts into the swing has no bearing on the time per swing because a more powerful motion will produce a wide arc and a higher speed of travel, whereas a low power swing will cause the weight to travel less distance at a reduced speed. Equally, the heaviness

25 Knight and Butler, *Before the Pyramids,* p, 31.
26 Knight and Butler, *Civilization One,* p. 34.

> of the weight of the object on the end of the line is immaterial — a heavier or lighter weight will simply change the speed/distance ratio without having any effect on the time of the swing.
>
> The mass of the Earth is a constant factor...

and thus

> in an area the size of the British Isles anyone swinging a pendulum for a known number of swings in a fixed period of time will have almost exactly *the same pendulum length.*[27]

So the method was simple: if one erected two markers on the circle of the horizon, spaced 1/366th of a degree apart, and then watching a selected star pass between them and, through trial and error with pendulum lengths, eventually a length would be found that would produce a half Megalithic Yard when swung 366 times as the star passed between the poles. And this would be the case *regardless* of where one was, and it would produce that length with *unerring accuracy.*[28] There was absolutely no need for an "ancient Bureau of Standards" whatsoever.

4. Beautiful Numbers: The 366-, 365-, and 360-Degree Systems

The 366-degree system also bears a close connection to a geodetic measure, namely, the polar circumference of the Earth. As Knight and Butler put it, "the most common value quoted" for the polar circumference of the Earth is "40,008 kilometers" which easily converts to "48,221,838 Megalithic Yards (MY)."[29] By assuming that these ancient Megalithic builders had divided each degree to 60 minutes of arc and each minute in turn to six seconds of arc, they are able to determine that one degree (or 1/366th) of the polar circumference of the Earth was 131,754 Megalithic Yards, and one minute (1/60th) of *that* value was 2,196 Megalithic yards. That, as they admitted, "did not look too exciting."[30] But then when one divided the minute by one second, or 1/6th of a part, the final result "was truly remarkable" for it yielded 366 Megalithic Yards![31]

This astonishing result tied into yet another ancient measure, this time the so-called "Minoan Foot" from the island of Crete. Noting that "Professor

27 Knight and Butler, *Civilization One,* p. 36, emphasis added.

28 Knight and Butler, *Civilization One,* p. 37.

29 Ibid., p. 29.

30 Ibid., p. 30.

31 Ibid.

J. Walter Graham of Princeton University" had made the discovery of "a standard unit of length" that had been used "in the design and construction of palaces on Crete dating from the Minoan period, circa 2000 B.C., Graham dubbed this unit a 'Minoan foot' which he stated was equal to 30.36 centimetres."[32] Knight and Butler then made the discovery that tied this measure to the whole 366-degree Megalithic system they had found in Britain:

> Imagine our surprise when we realized that one second of arc in the assumed Megalithic system (366 MY) is equal to 303.6577 metres — *which is exactly 1,000 Minoan feet* (given that Graham did not provide a level of accuracy greater than a tenth of a millimetre). This fit could just be a very, very strange coincidence — but it has to be noted that several researchers now believe that the Minoan culture of Crete had ongoing contact with the people who were the Megalithic builders of the British Isles.[33]

But this was not all.

There was in use in ancient Greece a unit of measure known as the "Olympian" or "geographical" foot, which, "by general consent" measures "what might at first seem like a meaningless 30.861 centimetres." And this raised even more peculiar relationships for Knight and Butler, for it exposed a hidden relationship between the Megalithic 366-degree system, and our now more familiar 360-degree system:

> We immediately noticed something special about the relationship between the Minoan foot and the later Greek foot. To an accuracy of an extremely close 99.99 percent, a distance of 366 Minoan feet is the same as 360 Greek feet! This was incredible, and we felt certain that it was not a coincidence. The two units did not need to have any integer relationship at all — yet they relate to each other *in a Megalithic to Sumerian manner...*[34]

In other words, Sumeria — which was the origin of our modern 360-degree system — had entered the picture, and in a way that connected the two systems, the 366- and 360-degree systems.

The link is evident from the Sumerian number system itself, which was a sexagesimal system, that is to say, based on units of sixes, tens, and six-

32 Ibid., p. 31.

33 Knight and Butler, *Civilization One,* p. 31.

34 Ibid., p. 137.

ties, and multiples thereof, thus making numbers such as 36, 360, 3600, and 36,000 very "Sumerian" numbers. As we have already noted, it appears that the ancients, in deriving the 366-degree system to begin with, had noticed the transit of the sun through a year was approximately 365.25 days, and simply rounded that number up to the next whole number, 366. The Sumerians, who influenced the Egyptians, modified or tempered the entire system by coming up with the 360-degree system in use to this day, to make calculation easier. In other words, at *first glance* it would appear that there was, at a certain point in prehistory, "Sumerian reform" to the Megalithic system. This is the position of Knight and Butler, namely, that the Megalithic system is the oldest, and the "Sumerian reform" a later modification of it. But as we shall see later in this chapter, there is evidence — from a very *unusual* place, that the 360-degree system was in use *long* before the Sumerians or Megalithic builders of Britain were even on the scene!

But whatever the *chronological* relationship between the two systems was, there was nonetheless a mathematical one, as was evidenced by the relationship of the Minoan to the Greek foot, the former representing a measure based on the 366 system, and the latter one based on the 360 system. The problem was, what was it?[35] Was it simply a case that "the 360-degree circle may have been a later adjustment to make arithmetic easier..."?[36] Or was something else at work in addition to this?

Knight and Butler quickly discovered what the mathematical relationship was, and the relationship was that of two absolutely critical and fundamental constants, those of f, with a value of 1.618, and π, with the value 3.14. Simply put, 360 divided by 5 equaled 72, and 366 divided by $\pi \times f$ *also* equaled 72. While the division by 5 may seem arbitrary, it is not, for the constant f generates the well-known Fibonacci sequence — 1, 2, 3, 5, 8, 13 and so on — where the first two numbers sum to the next number, then those two to the next, and so on. So five is in the "harmonic series," so to speak, of f.

Then Knight and Butler made another astonishing connection between the two systems. Noting that the Thornborough henges pointed to the mount in Lincoln over a distance 127 kilometers away, and finding that the two sites lay exactly one Megalithic degree apart, that is, 366 x 60 x 6 Megalithic Yards from each other, *by latitude* and exactly *one* degree apart by the modern system of longitude meant, in effect, that for a period of time the Megalithic builders appear to have used the 366 to measure longitude and the 360-degree system to measure latitude.[37] The "Sumerian reform" on which our

35 Knight and Butler, *Civilization One,* p. 131.

36 Ibid., p. 27.

37 Knight and Butler, *Civilization One,* p. 106.

entire modern system is based, in other words, had consisted of nothing but the use of the 360-degree system to measure both.

5. The Next Step: Measures of Volume and Weight

Having thus found the means whereby an accurate unit of linear measure was so accurately reproduced over so wide an area — namely, through the use of a pendulum to count beats between markers on the circle of the horizon placed one degree apart in a 366-degree system — and thus converting the measure of *time* to a linear measure of *distance,* the next problem was to introduce accurate units of measure of volume and weight *based on that same linear measure.* "Such a move would have been an important building block *towards trade, which was in turn a key step towards true civilization.*"[38] In other words, the necessary basis for international trade is accurate units of measure of volume and weight, and that accuracy can only be guaranteed by basing units of measure on the relative constancy of celestial motions and the geodetic properties of the Earth itself.

The problem was easily solved by performing what physicists call a "dimensional rotation," i.e., simply taking the linear measure, a unit of *one* dimension, and rotating it into *two* dimensions (thus measuring *area*), and then finally, into *three* dimensions, thus measuring volume. In short, the ancients simply "cubed" their Megalithic Yards, half-yards, and so on, to form the measures of volume.

But when this was done, Knight and Butler began to make another astonishing series of discoveries.

> In our case, the linear units would have to be in Megalithic Inches, which Thom identified as being one fortieth of a Megalithic Yard, equal to 2.07415 centimetres. Taking his lead from the metric system Chris first considered a cube with sides of a tenth of a Megalithic Yard — i.e., four Megalithic Inches (MI). *In metric terms this turned out to have a capacity of a little over half a litre, at 571.08 cubic centimetres.*[39]

This result — an odd parallel between an ancient Megalithic system of measure and the modern metric one — was odd, but hardly compelling in its own right.

38 Ibid., p. 47.

39 Knight and Butler, *Civilization One,* p. 53.

6. Ancient and Megalithic Anticipations of the Imperial and Metric Systems

But then the problem became acute, and was disclosed by a conversation between co-authors Alan Butler and Chris Knight. I reproduce it here, because their reactions would most certainly have been my own, and most likely anyone else's doing the experiments and running the numbers. The conversation begins with author Chris Knight:

> "...I think we have a problem."
>
> "What sort of problem?" Alan wanted to know.
>
> "The problem of explaining the apparently impossible," said Chris. "I started by checking out *spheres* with diameters of 5, 10 and 20 Megalithic Inches and *they also produce volumes that are quite close to the pint, one gallon, and the bushel.* The accuracy level isn't quite as good as the cubes because the 5 (megalithic inches) sphere held 1.027 pints which is still as close as anyone in the real world would ever need. But a quick check of the rules that govern the relationships between cubes and spheres revealed that to an accuracy of 99.256 percent a cube with a side of 4 units will have the same volume as a sphere with a diameter of 5 units, which made the findings odd but mathematically understandable."
>
> Alan was intrigued but puzzled.
>
> "If there is no mystery about the pint sphere, why did you say you had to explain the impossible?" he asked.
>
> "What I've told you so far is the easy part of this conversation, because my next test took me from the rather weird to the downright ridiculous. What do you think that a 6 MI and a 60 MI diameter sphere would hold in terms of weight of water?"
>
> "I can't guess. What do they hold?" Alan asked, with not a little impatience.
>
> "Well, the 6 MI sphere holds a litre and weighs a kilo, so the 60 MI sphere 10 x 10 x 10 times as much, holds a cubic metre and weighs a metric tonne. And it's incredibly accurate too."
>
> Alan laughed aloud down the phone.
>
> "Ha ha, very funny..." He paused. "You are joking, aren't you?"
>
> "No. You check it out, Alan. The numbers don't lie. The fit is better than 99 percent accurate and when I tested the same principles using modern inches and centimetres for the spheres, there were no meaningful results at all. Something truly weird is going on here."[40]

40 Knight and Butler, *Civilization One,* pp. 56–57, emphasis added.

In other words, when modern imperial inches or metric centimeters were used to construct spheres and cubes, there were "no meaningful results at all." But when the ancient Megalithic inch was used to construct cubes and spheres, the result was, in the case of the cubes, modern British imperial units of volume, and in the case of the spheres, modern metric units of volume and weight! In other words, *both the modern imperial and metric systems of volume appeared to be related to the ancient units of measure precisely as being based on cubing and sphering.* And further research revealed that even the Imperial weights appeared to be derived by cubing a unit one-tenth the length of the Megalithic Yard.[41]

But in order to demonstrate the suggestion that there may indeed be an *actual* link between the two systems, a link stemming from ancient times, one would have to reproduce an ancient unit of measure that bears a resemblance to one or the other system. There was indeed such a unit, which the Sumerians called the "double kush," approximately 99.88 centimeters long, which, oddly enough, made the length of a pendulum of one second for accurate timekeeping.[42]

All of this was compelling evidence that there was not only a deeply ancient root and relationship between the modern British imperial and metric systems of volumes and weights, but that that root and relationship stemmed from a unit of Megalithic linear measure that was cubed and "sphered." All of this suggested, in other words, as Knight and Butler themselves commented, that someone very much wanted to "jump-start" civilization and create the conditions necessary for "international trade," someone like a hidden elite.

And this is where the story of that hidden elite, and its hidden agendas, takes another astonishing turn.

B. The Hidden Elite and the Cosmic War Scenario

1. *The Ancient Elite: Astronomy, Finance, and the God of Corn Versus the God of Debt*

Knight and Butler were not blind to the implications of their discoveries, and posited the existence of such a hidden elite, a guiding hand, steering Megalithic man inevitably and inexorably along the road to (or, as we shall see, *back* to) civilization. How indeed was it possible "that the supposedly un-

41 Ibid., p. 59. Knight and Butler also speculate on the existence of a Megalithic Rod that is 6 Megalithic Yards in length, or 5 modern meters in length. This gives a unit of linear measure that relates the imperial and metric systems directly, since a modern statutory mile would then be equal to 320 "Megalithic Rods" in length, while a modern metric kilometer would equal 200 Megalithic Rods in length. See pp. 62–63 for these results.

42 Knight and Butler, *Before the Pyramids,* p. 78.

sophisticated people of Stone Age Britain possessed a fully integrated system of measurement based on a deep understanding of the solar system?"[43]

As they were investigating the British sites where Thom discovered the Megalithic Yard, Knight and Butler soon found an answer to that question, and it was an answer with deeply significant implications. For example, the fact that there were so many such Megalithic sites scattered throughout the British Isles, and that each seemed to be designed for some astronomical purpose, suggested that "there might have been a national *network*"[44] of some sort in place. And a network implies the organization and coordination of a select group of people. In short, it implies an elite.

One such site in particular intrigued Knight and Butler, and this was at Skara Brae in Orkney Island. It intrigued them because

> it may well have been a Megalithic 'university' for training astronomer-priests. Radiocarbon dating has shown that it was occupied between approximately 3215–2655 B.C. when it provided a series of linked rooms, each with matching stone-built furniture including dressers, beds, cooking areas and sealed stone water tubs for washing. Archaeologists have identified that secrecy, security and plumbing are also apparent at the site. A secret hidey-hole had been found under the stone dresser and a hole for a locking bar was located on both sides of doors. In addition, a lavatory drain designed to run excrement along wooden piping and into the sea has also been excavated....
>
> .... *Because the island had nothing to trade, the only reasonable answer to this archaeological puzzle is that the inhabitants had been an elite group who were supported by the goodwill of a broader community at a distance.*
>
> Skara Brae also revealed some artifacts that have proved impossible to understand. Small stone objects that have been exquisitely carved include two balls: one 6.2 centimeters and the other 7.7 centimeters in diameter. *Their purpose is unknown and the deep decoration appears to be impossible to create without metal tools* as engineer James Macauley discovered when he attempted to reproduce them using the known technology of the time.[45]

In other words, Skara Brae was basically confirmation of the existence of a supported elite, one that moreover had access to *some* sort of advanced metal and stone-working technology that was *not* common to the wider society supporting it.

---

43 Knight and Butler, *Civilization One,* p. 1.

44 Knight and Butler, *Civilization One,* p. 52.

45 Ibid., pp. 52–53, emphasis added.

Skara Brae is also illuminating for another reason, one that Knight and Butler do not mention, for if there was an elite guiding a post-cosmic war humanity back up the long ladder to civilization, two of the essential steps that it would take to do so are (1) to establish a system of weights and measures accurately reproducible over time and at any point on earth by means of relatively simple methods, and (2) it would establish a center or centers to train people in the proper techniques of deriving these measures via astronomical observation.

Skara Brae, in other words, is testimony not only to the existence of that elite, but also to its *agendas,* for

> Without a means of gauging weight and volume, trade remains at a bartering level where each transaction has to be assessed by visual evaluation alone. The ability to identify a known quantity makes buying and selling a much more scientific process since it can be accurately repeated time after time. Using mutually accepted units of weight meant deals could be done at long range because it would be unnecessary to see the merchandise first to assess its quantity.[46]

And thus we are in the presence of yet another agenda of that elite, for so long as there was no accurate system of measure that could easily and simply be reproduced *anywhere on earth,* what trade and civilization there was would be confined to very small local areas. Hence, the presence of a "Megalithic university" suggests very strongly that we are in the presence of an elite that is attempting to foster trade over the widest area possible in order to create the necessary and fundamental conditions for the emergence of a global civilization. We are thus chin to chin with that deeply ancient connection between the hidden financial elites of ancient times, and the astronomy-temple priesthoods that I referred to in my previous book, *Babylon's Banksters,*[47] for that alliance was present even prior to the emergence of the high civilizations of Sumer and Egypt. That elite, with its alliance of astronomy with a hidden financial and political agenda, indeed *created* those civilizations.

A brief word about that Megalithic financial-astronomical priesthood is in order, lest its place within the wider context of my previous books be misunderstood. As I noted in *Babylon's Banksters,* the initial and earliest idea of money was that it was simply a unit of exchange issued by the state itself against the surplus of production in the state warehouses. As that production

46 Knight and Butler, *Civilization One,* p. 67.

47 See my *Babylon's Banksters: The Alchemy of High Finance, Deep Physics, and Ancient Religion* (Port Townsend, Washington: Feral House, 2010), pp. 159–205.

expanded, the money supply did; as it contracted, the money supply did. But the crucial point to observe is that money was issued *as a receipt on the production of the state, and thus was issued debt-free.* Since the probity of the state was represented precisely by that astronomical-financial priesthood that had indeed called it into being, the issuance of "legal tender" — to use the modern parlance — was often associated with the temple. Thus, initially, that astronomical-financial elite subscribed to a monetary policy that would, in terms of religious imagery, be described as a worship of the "God of production," the "Corn God."

However, as I also detailed in *Babylon's Banksters,* that elite was soon infiltrated by another, or co-opted, and its own members corrupted from this initial "Corn God worship and monetary policy" and turned to a philosophy of the *private monopoly issuance of debt as money.* Or to put the change in monetary policy in terms of the religious imagery of the temple once again, worship turned from the Corn God to the God of Contract and Debt, to the God to Whom a debt was owed and service *due.* To view the monetary policy in such religious terms is to reveal the dirty little secret that began to take hold on human culture as a result of that change in the elite that began to occur in Sumeria and Babylonia, for as anyone knows, banks can only issue *principal,* not *interest,* and thus as the circulation of monetized debt grows and replaces the circulation of real money, the debt only increases for the many, to the profit of the few at the top of the pyramid and Ponzi scheme. Or, to put it in its ultimate religious expression, mankind comes by fits and starts into the inevitable situation where an infinite, or at least practically un-repayable, debt is owed to God, or the gods, requiring an infinite treasury to repay.[48] To put it as succinctly as possible, the two monetary policies manifest themselves as the spiritual consequences of the worship of two very different Gods, and vice versa.[49]

We have, then, that initial elite, and then another one that snakes its way *into* the original, diverting its monetary policy, while retaining many of the original agendas, including that of the creation of a global civilization.

2. The Masonic Elite and the Lore of "Very High Antiquity"

The presence of the Megalithic measures suggested, as has been seen, an elite with an agenda, an agenda which included its own self-preservation and

48 The apogee of this sort of monetized debt-as-spirituality was reached, of course, in the Latin Middle Ages with Anselm of Canterbury's *Cur Deus Homo,* which might best be described as a study of the monetized debt of man to God.

49 And this point *might* be a possible key to disentangling the layers of various traditions within various texts.

continuance over time. Knight and Butler put the point this way:

> Indeed, the fact that the Megalithic units have an almost perfect relationship with modern measurements strongly suggests that there has been a continuity of this knowledge across the Great Wall of History.... That brought us to the inventors of writing and the first known nation of international traders, the Sumerians...[50]

The Great Wall of History to which Knight and Butler refer is the "wall of silence" between the Megalithic Age and the earliest civilizations. But as we shall see, that wall is considerably higher and larger than they imagine. Its import, however, is the same: what relationship was there between the Megalithic builders and their measures, and Sumeria? Clearly there was a persistence of information across that Great Wall of History, implying that an elite had carried that information forward.

As Knight and Butler themselves point out, however, they were not the only ones to notice the implications; the other was none other than America's third president, Freemason Thomas Jefferson. Jefferson investigated systems of weights and measures shortly after America gained political independence from Great Britain, and like Knight and Butler, made a similar series of discoveries about weights and measures in use in Britain and France as did Knight and Butler, and these discoveries led him to some rather astonishing conclusions:

> What circumstances of the times, or purposes of barter or commerce, called for this combination of weights and measures, with the subjects to be exchanged or purchased, are not now to be ascertained. But a triple set of exact proportionals representing weights, measures, and the things to be weighed and measured, and a relation so integral between weights and solid measures, must have been the result of design and scientific calculation, and not a mere coincidence of hazard.
>
> ....But the harmony here developed in the system of weights and measures... corroborated by a general use, *from very high antiquity*, of that, or of a nearly similar weight under another name, seem stronger proofs that this is legal weight...[51]

Knight and Butler decided to follow Jefferson's lead and see if his comment about "very high antiquity" had any possibility of truth.

---

50 Knight and Butler, *Civilization One,* p. 68.

51 Cited in Knight and Butler, *Civilization One,* p. 108.

Taking their cue from an ancient Sumerian tradition that spoke of the world being measured in barley seeds,[52] Knight and Butler decided to see if there was any possible truth to the Sumerian tradition. In doing so, they discovered the ultimate basis of the Sumerian sexagesimal numerical system: the mass of the Earth itself!

Since the Sumerian tradition had stated that the weight of the Earth was measured in barley seeds, Knight and Butler began with the standard value of the metric mass of the Earth, 5/9763 x 10^{24} kilograms. They then converted this figure into a Sumerian unit of weight based on cubing the Sumerian double-kush. Once cubed, and filled with water, this in turn became the Sumerian unit of mass called the "double-mana," which weighed 996.4 grams. This meant that there were 5.9979 x 10^{24} double-manas in the Earth's mass.[53]

Their commentary on the astonishing nature of what follows must now be cited to allow its full significance to sink in:

> This number is as close to 6 followed by 24 zeros as to stand out as being very odd indeed, particularly bearing in mind that we could not be certain as to the "exact" intended size of the double-kush. Of course, it could be a coincidence but it remains a fact that the weight of the world is only one part out in 2,850 from being precisely:
>
> 6,000,000,000,000,000,000,000,000 Sumerian double-manas.
>
> If it were not for the fact that this number conforms so spectacularly to the Sumerian/Babylonian base 60 system of counting we would not have reported it. But it is a tantalizing thought that this ancient unit may have a relationship to the mass of the Earth, either by some brilliant calculation or due to some practical experiment that produced the result by a mechanism unknown to the originators — or to the modern world. Furthermore, we knew that the Sumerians considered that there were 21,600 barley seeds to one double-mana so we can also venture to say that the entire planet is equal to 1,296 x 10^{26} barley seeds — which then gives the following result:
>
> One degree slice of the Earth = 360 x 10^{24} barley seeds
> One minute slice of the Earth = 6 x 10^{24} barley seeds
> One second slice of the Earth = 10^{23} barley seeds.

52 Q.v. *Civilization One*, pp. 115–120.

53 Knight and Butler, *Civilization One*, p. 121.

> So, a one-second-wide section of our planet weights the same as an incredibly neat 100,000,000,000,000,000,000,000 barley seeds. Simply astonishing![54]

But the problem was just beginning!

Having already noticed the basis of the imperial and metric systems in the Megalithic measures, Knight and Butler decided to try a similar experiment for the mass of the Earth in the imperial system. When this was done, they obtained the following results:

1 Megalithic-degree section of the Earth	$= 360 \times 10^{20}$ pounds
1 Megalithic-minute section of the Earth	$= 6 \times 10^{20}$ pounds
1 Megalithic-second section of the Earth	$= 10^{20}$ pounds[55]

Their conclusion pointed inevitably to the existence of *something before* Sumeria, and to a hidden elite that had passed on these units of celestial and geodetic measure:

> This could still be a double, outrageous coincidence but the odds against both systems fitting like a near-perfect glove and bearing in mind the Sumerian base 60 method of calculation, made it seem impossible. Somebody in the distant past appears to have known the mass of the Earth to a very accurate number.
>
> ... The relationship of the pound weight and the double-mana (virtually a kilogram) to the mass of the Earth did not seem compatible with the level of sophistication of either the Megalithic people or the Sumerians. Could some other unknown group have developed the principles we see in use and then taught them to these fledgling cultures? Is mankind's leap across the Great Wall of History due to some super-culture that has left no other trace of itself? For the first time we began to theorize about the strange possibility of a group whose existence can only be deduced by the knowledge they left behind.[56]

---

54 Knight and Butler, *Civilization One,* p. 122.

55 Ibid., p. 124.

56 Knight and Butler, *Civilization One,* p. 125.

But if civilizations such as Sumeria and Egypt were able to produce such systems and maintain them with accuracy, would it not be inconsistent to ignore what they themselves say about their own origins? And they say, indeed, that at the first stages of their development — a development stretching far beyond the Megalithic builders and into hundreds of millennia prior to that according to the Sumerian Kings List[57] — that they were indeed founded by "god-kings" who taught them the arts of civilization, arts that would have included accurate systems of weights and measures. Moreover, the sheer *extent* of the presence of such measures, from Britain to Sumeria, would also suggest that we are in the presence of an elite that is scattered — or that has intentionally scattered itself — over the surface of a wide area of the Earth as a kind of "paleoancient international" priesthood of trader-astronomers.

C. The Cosmic War: Marduk Measured the Structure of the Deep

So how does all this measuring activity in ancient times fit into the "Cosmic War" scenario? What agendas might it disclose beyond that already suggested, namely, that an elite was in a hurry to stimulate trade based on reliable and accurately reproducible units of measure based on astronomical observation?

The answer to these questions is suggested by a short statement that occurs at the end of the fourth tablet of the so-called Babylonian creation epic, the *Enuma Elish.* The statement is that "The Lord," i.e., the Babylonian god Marduk, "measured the structure of the Deep."[58] But the context reveals why this was necessary, and what is meant within the context by "the Deep":

> 47. (Marduk) sent forth the winds which he had created, the seven of them;
> 48. to disturb the inward parts of Tiamat, they followed after him.
> 49. Then the Lord raised the thunderbolt, his mighty weapon,
> 50. He mounted the chariot, the storm unequalled for terror,
> 51. He harnessed and yoked it unto four horses,
> 52. Destructive, ferocious, overwhelming, and swift of pace...[59]
> 58. With overwhelming brightness his head was crowned...
> ...
> 65. And the Lord drew nigh, he gazed upon the inward parts of Tiamat...[60]

57 Q.v. my *The Cosmic War: Interplanetary Warfare, Modern Physics, and Ancient Texts,* pp. 191–203.

58 *Enuma Elish,* ed. L.W. King, M.A., F.S.A., Vol. I (London: Luzac and Co., 1902), Tablet 4, p. 77.

59 Ibid., Tablet 4, p. 65.

60 Ibid., Tablet 4, p. 67.

75. Then the Lord (raised) the thunderbolt, his mighty weapon...
76. (and against) Tiamat, who was raging, thus he sent (the word):
77. "(Thou art become great, thou hast exalted thyself on high,
78. and thy (heart hath prompted) thee to call to battle..."[61]

...

87. When Tiamat heard these words,
88. She was like one possessed, she lost her reason
89. Tiamat uttered wild piercing cries,
90. she trembled and shook to her very foundations....

...

95. The Lord spread out his net and caught her,
96. and the evil wind that was behind him he let loose in her face.
97. As Tiamat opened her mouth to its full extent,
98. He drove in the evil wind, while as yet she had not shut her lips.
99. The terrible winds filled her body...

...

101. He seized the spear and burst her body,
102. He severed her inward parts, he pierced (her) heart.

...

129. And the Lord stood upon Tiamat's hinder parts,
130. and with his merciless club he smashed her skull.[62]

...

137. He split her up like a fish into two halves....

...

143. And the Lord measured the structure of the Deep.[63]

These verses, in my opinion, speak less of a "creation epic" based on a dualistic cosmology, a dualism allegorized as a war (the standard academic interpretation), as they do of an *actual* war, where "Tiamat" stands as the name of an actual planet that once existed within our solar system. Thus viewed, the verses are a garbled, though nonetheless clear indication of a sophisticated technology and hence of a sophisticated civilization with a very sophisticated physics. I have stated this interpretation in the following fashion:

> I believe these passages reveal a remarkably accurate sequence of what the destruction of a planet by a "scalar" weapon employing a longitudinal pulse or acoustic stress in the medium itself would entail,

61 Ibid., Tablet 4, p. 69.
62 *Enuma Elish,* Tablet 4, p. 71.
63 Ibid., p. 73.

right down to acoustic cavitation and large electrostatic displays, signatures of the use of such a weapon at extreme power. Let us note the sequence:

(a) The "winds" are sent to "disturb" or destabilize the "inward parts" of Tiamat, the planetary core (vv. 47–48);
(b) "Lightning" is then unleashed on the (already destabilized) planet from the "four winds," i.e., from every direction (vv. 49–50)....
(c) These "thunderbolts" are then apparently directed toward that destabilized core, suggesting that a sudden and extreme *pulse* is administered (vv. 58, 65, 75–78);
(d) Tiamat responds with cries and trembles and shakes to "her very foundations," i.e., experiences very severe earthquakes or acoustic cavitations throughout the planet, to its very core (vv. 95, 97);
(e) Tiamat appears unable to break resonance with the weapon (vv. 97–98) as Marduk spreads the net and drives in the final "wind" or pulse (v. 98);
(f) Tiamat reaches maximum instability in her planetary core and mantle (cc. 98–99);
(g) Marduk pierces the crust, and releases the enormous energies that have built up in the planet through the acoustic cavitations, resulting in a colossal explosion with the entire planet as its fuel, rather like bursting a balloon filled to extreme pressure (vv. 101–102, 137).

All this implies the existence of a physics sophisticated enough to "measure the structure of the Deep" (v. 143), and to weaponize it...

Another comment is perhaps warranted by this discussion. It is to be noted that Marduk "measures the structure of the Deep" *after* Tiamat's destruction. This would have been necessary in terms of the type of physics *being suggested, since the destruction of a planetary-sized body in the approximate orbit of the asteroid belt would have required an adjustment to astronomical measurements of the solar system, since its previously existing celestial mechanics and geometry has been shattered.*[64]

To summarize what is being argued in the Cosmic War context, we have the following:

---

64 Joseph P. Farrell, *The Giza Death Star Destroyed* (Kempton, Illinois: Adventures Unlimited Press, 2005), pp. 46–48.

1) A physics once existed whereby it was possible to tap into the geometries of the local physical medium to the extent that it was possible to blow up an entire planet in an act of war;
2) That physics could only be accessed by accurate *measures* of "the Deep" or of local space and its celestial mechanics;
3. After the destruction caused by that Cosmic War, it was necessary for the surviving elites to quickly re-establish accurate measures of the surviving celestial mechanics for *two* reasons:

    a) If civilization were ever to return to a similar state of development with access to that physics once again, of necessity that civilization would have to be global in extent, and this could only be achieved through the gradual re-establishment of global trade, which in turn depended on accurate, uniform, and consistent units of measure, which in turn could only be established by "measuring the structure of the Deep," and this is precisely what we have seen happened in the Megalithic measures.
    b) If the technology that made such fearsome weapons possible were ever to be reconstructed, again, this would require accurate measures of "the structure of the Deep."

Thus, we are perhaps in the presence of *two* hidden long-term agendas, and very possibly in the presence of *two* elites with vastly different long-term objectives, with one elite wanting to restore the same level of civilization as existed before the war, and the other wanting to reestablish the fearsome technologies that made that war possible, and thereby to restore, or claim, its own hegemony. At the minimum, then, the Megalithic measures and their self-evident propagation by a hidden elite — as Knight and Butler have shown — is a demonstration of the *first* agenda. As we shall discover in the coming pages, however, there is ample evidence to suggest that *another* elite is at work for purposes that are not so benign. And this brings us back to Sumeria, and the suggestions of that deeper physics.

1. The Suggestions of a Deeper Physics

As has been seen, the principle basis of the ancient units of measure was the conversion of a regular measure of time into a unit of linear measure via a simple "pendulum method." Then by the process of cubing or "sphering" that linear measure, that unit of linear measure was converted into units of measure for volume and, when those cubes or boxes were filled with a quantity of a known substance, such as water, units of measure for mass, thus reproducing simply and elegantly accurate units of measure based on known astronomical phenomena invariant over the surface of the Earth.

But Knight and Butler did not let things stand there, for the observation that time and linear measure were deeply and intimately connected was, of course, one of the profound insights of Albert Einstein and his special theory of relativity. Sensing that the ancient measures possibly pointed to knowledge of a much deeper physics, they decided to subject the Sumerian measures to "the acid test" to determine if, in fact, their knowledge might have come from a much more ancient, and much more sophisticated, civilization:

> Throughout our investigation we have tried not to prejudge what is, and is not, possible for an ancient culture to achieve. We have simply tried to let the data lead us to wherever it takes us. But at this point we were starting to get cold feet. *We seemed to be uncovering complexities that surely must have come from a highly developed society with advanced scientific abilities. With this uncomfortable thought in our minds we decided to try the most obvious next experiment involving the most fundamental property of the universe — the speed of light.*
>
> Could the Sumerians possibly have understood how fast light travels? According to current knowledge light travels at 299,792,458 meters per second in a vacuum, which translates to Sumerian units as 600,305,283 kush.
>
> ....
>
> We decided to look at what is known of the speed of our own planet as it orbits the Sun and found that the near-perfect circle of the Earth's path is 938,900,000,000 metres, which is covered in a year of 365.2596425 days. These numbers look remarkably unimpressive but the next calculation left us staring at the calculator in disbelief. We were stunned to find that we all travel on our yearly journey at speed of 60,000 kush per second. As a further level of strangeness this speed is a neat one-ten-thousandth of the speed of light.
>
> The standard response of mathematicians to numbers that look incredibly neat is to yawn, because they believe that all numbers are equally probable and the actual digits are dependent on the numerical base and the measurement convention employed. They are quite right. But they assume that all measurement units are merely a convention without any underlying physical reality. And that is not the case with either the Megalithic or the Mesopotamian systems.[65]

65 Knight and Butler, *Civilization One,* pp. 126–128, emphasis added.

Their conclusion was as inevitable as it was astonishing:

> In this case the second and the kush appear to be very much more than a convenient abstraction because they have all of the characteristics of being fundamental to the realities of the earth's environment. They have a value at a level never conceived by modern science. *We have come to the conclusion that it is more than reasonable to believe that the Sumerians, or more probably their unknown teachers, understood both the mass of the Earth, its orbital speed and even the speed of light, and they designed units that had an integer relationship with them all.*[66]

To say that this was the case — that there was a pre-existent civilization *seeding* its knowledge into Sumerian civilization and culture — is to say that there is a deep physics to its units of measure and an elite that propagated them.

To see why, we need but recall the fact that these units of measure were based on "the mass and spin of the Earth,"[67] and any time one couples the idea of *mass* and *rotation,* one is dealing perforce with the concept of *torsion.* Torsion has been a "physics theme" of many of my books,[68] but for our purposes here we may understand torsion as the spiraling, folding, and pleating of the fabric of space-time around *any* rotating mass. If one wishes to draw a mental picture of what torsion accomplishes, the analogy of wringing an empty soda can like a dishrag is helpful. As one wrings the can, the can spirals and folds and pleats, and the ends of the can draw closer together. In this analogy, the can would represent space-time.

2. The "Sumerian" Mysteries of Deep Space

But is there any other confirmation that Knight's and Butler's conclusion — that the Sumerians derived their knowledge from someone else, someone much more ancient and much more advanced — than merely numerical coincidence? Indeed there is, and it comes from renowned space anomalies investigator Richard C. Hoagland, or rather, from some very remarkable NASA photographs of an object doing some very "Sumerian" things.

That object is Saturn's "moon," Iapetus and the remarkable — nay, mind-boggling and unbelievable — photographs of it taken by NASA's probe

66 Ibid., p. 128, emphasis added.

67 Ibid., p. 2.

68 See, for example, *Secrets of the Unified Field: The Philadelphia Experiment, the Nazi Bell, and the Discarded Theory* (Adventures Unlimited Press, 2008), pp. 1–42, 172–190, 248–252, 262–288, and *The Philosophers' Stone: Alchemy and the Secret Research for Exotic Matter* (Feral House, 2009), especially pp. 151–200, 313–326.

Cassini. In what is undoubtedly one of his most fascinating pieces of analysis and discussion of NASA images, Hoagland reproduced the following *Cassini* images in a paper entitled "A Moon with a View." As will be evident to anyone who looks at the pictures, Saturn's "moon" Iapetus is not a moon at all, it's a nine-hundred-mile-wide artificial body.

The first testament of that fact is that Iapetus, unlike "moons," has extraordinarily *straight edges* when viewed in highlighted by the light of the Sun:

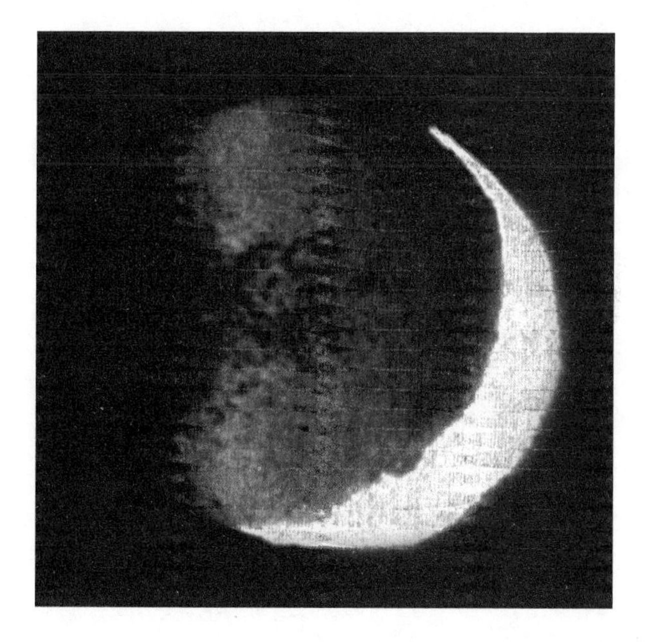

Iapetus' Straight Edges

And in close-up, these are even more evident:

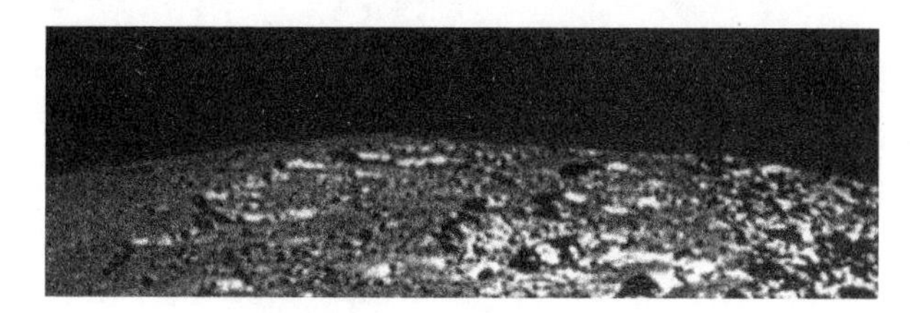

Iapetus' Straight Edges in Close-Up

But that is not the only difficulty. The *really* major problems are the three *parallel ridges* that run the circumference of its equator, a geological impossibility:

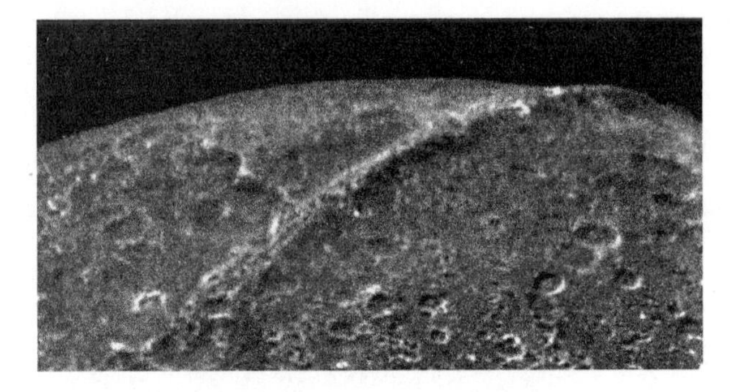

Iapetus' Equatorial Parallel Ridges

As if that's not enough, when viewed full-on, Iapetus displays a large hexagonally-shaped "crater" just above these equatorial "ridges":

Iapetus' Hexagonal "Crater"

If that picture looks hauntingly familiar, that's because it is, and Hoagland was quick to spot its resemblance to something much more sinister in recent popular culture: George Lucas' "Death Star" from his famous *Star Wars* movie epics:

Richard C. Hoagland's Iapetus-Death Star Comparison[69]

None of this, however, even comes close to comparing with what Iapetus is *doing,* and *how* it is doing it.

As Hoagland notes in his paper, Iapetus is doing some very odd things as it orbits Saturn. In fact, the odd things *are* its orbit around Saturn. Like Earth's own massive moon, Iapetus's orbit around Saturn is a nearly perfect circle. And also like Earth's moon, Iapetus turns on its own axis in such a fashion that the same hemisphere is always facing Saturn! And unlike *all* of Saturn's other moons, Iapetus is inclined 15 degrees from its equator. But that's not all. It's best to allow Hoagland to speak for himself here:

> Once the (amazing) possibility is admitted that Iapetus could be an *artificial* "moon" — and may have been deliberately *inserted* into such an odd orbit — the "coincidental" nature of its unique, steep inclination (relative to the other similar-sized Saturnian moons — Dione, Rhea, etc.) goes away.
>
> But, equally "coincidental" is the shape of Iapetus' inclined orbit...and its precise *distance* from Saturn. Iapetus' orbit is extremely close to being circular — with an eccentricity of only 0.0283 departing from a perfect circle by slightly less than 3%. (By comparison,

69 Richard C. Hoagland, "A Moon with a View," www.enterprisemission.com. Hoagland observes, rightly, that such a stark and striking resemblance challenges the notions of coincidence and synchronicity, as it raises the prospect of a "controlled release" of information, and raising the question of "how did Lucas know, when did he know it, and who told him?" Of course, such ideas may easily have been out of the control of Lucas himself, and may have come from someone in his production staff. But the larger issues remain: who knew, and how did they know long before *Cassini* took its breathtaking pictures? One answer, of course, is that the "moon" was secretly photographed long before *Cassini* journeyed to Saturn, which raises the prospect of a *secret* space program, a hypothesis that Hoagland has been advocating strongly for years.

> our Moon's orbital eccentricity... is 0.0549 or ~6%...essentially *twice* as eccentric as Iapetus!)
>
> For an almost *circular,* very high inclination orbit to have formed through "random chance" is really pushing coincidence — if the agent for achieving that low eccentricity *and* the high inclination is supposed to be the same *"random"* collisional event, back when Iapetus was forming.[70]

As I have noted elsewhere, such orbital mechanics in the case of our own Moon has led some scientists, particularly in Russia, to have concluded that the Earth's own satellite exists in an artificial and deliberately created orbit, as it has also led some to argue that our own Moon may *be* an artificial body.[71]

But, as Hoagland quips, "there's more":

> The sharp reader will have noticed, from the preceding references, that Iapetus orbits slightly less than 60 radii away from Saturn (59.09) radii, to be exact...). This discrepancy, 0.15% — in the artificial model that *precisely* 60 radii was originally intended — would represent how much Iapetus has drifted since it was "parked" (as a designed "station") in Saturn's orbit. That rate of drift, either due to Saturnian/sun tides, or other forces...could give another way to estimate — other than by counting craters — roughly "when" this entire scenario in fact occurred....[72]

As I noted elsewhere, what Hoagland is referring to is the fact "that Iapetus orbits Saturn some two million miles from the planet, which is almost exactly sixty times the radius of Saturn."[73] By now that number — sixty — should sound very familiar, for it is the whole *basis* of the Sumerian sexagesimal numerical system! Hoagland is quick to see this, for "that 'ideal' Iapetus distance from Saturn just 'happens' to also be base 60... suddenly appearing *in the first Sumerian civilization on Earth some 6000 years ago...*"[74] To put it as succinctly as possible: it would appear that someone had "parked" Iapetus in an orbit around Saturn using the very basis of a sexagesimal system that only emerged on Earth sometime later, and thus, that there *may indeed be*

70 Richard C. Hoagland, "A Moon With a View," part four, www.enterprisemission.com, emphasis Hoagland's.

71 See my *The Cosmic War: Interplanetary Warfare, Modern Physics, and Ancient Texts* (Adventures Unlimited Press, 2007), pp. 337–359.

72 Hoagland, op. cit.

73 Farrell, *The Cosmic War,* p. 389.

74 Hoagland, op. cit., emphasis added.

a cultural connection between Sumeria and an artificial body in our solar system millions of miles from the Earth!

3. A Chronological Conundrum

All this places the origin of Megalithic measures, and the probable role of an elite in deriving them, into an entirely different context, and it raises profound and thorny chronological questions. As I outlined in my previous book *The Cosmic War,* if such an interplanetary war did occur, it most likely occurred ca. 3.2 million years ago.[75] But the measuring activity represented by the Megalithic builders at best represents an activity that was begun no earlier than 10,000–12,000 years ago,[76] and in the case of the British Megalithic builders, an activity only a few thousand or so years older than the Great Pyramid.[77] This places the chronological problem into stark relief, for presumably such measuring activity would have begun shortly after the conclusion of the war, yet the evidence suggests it was begun some three million years later. While there is *no* easy way around this problem, it is possible that such activity may have begun when those surviving elites thought the population of the Earth had been restored to a sufficient level to warrant their activities in creating the necessities of civilization. Nonetheless, the problem remains, and there may be as yet undiscovered facts that would fill in this lengthy gap more plausibly.

D. Conclusions

So what may be concluded from the existence of the Megalithic measures and their incredible basis in accurate astronomical measurement? What may be concluded from the presence of Sumerian measures in the orbital mechanics of Saturn's "moon" Iapetus? What agenda or agendas do these dimly disclose? From what has been examined thus far we may comfortably conclude the following:

---

75 Joseph P. Farrell, *The Cosmic War: Interplanetary Warfare, Modern Physics, and Ancient Texts,* pp. 132–133, 199–203.

76 See the important work by Thomas G. Brophy, Ph.D., *The Origin Map: Discovery of a Prehistoric, Megalithic, Astrophysical Map and Sculpture of the Universe* (New York: Writers Club Press, 2002). Brophy's work documents in meticulous detail that the ruins of Nabta Playa in Egypt actually constitute an accurate astronomical depiction of the Milky Way Galaxy, dating from ca. 10,000 B.C.

77 Q.v. Knight and Butler, *Civilization One,* pp. 12–14. Of course, this view holds that the Great Pyramid is a much more recent structure than Nabta Playa, a view that the current author does not share. Knight and Butler also argue that the layout of the Giza plateau was actually planned and determined in a series of megalithic sites in Great Britain much earlier. Q.v. Knight and Butler, *Before the Pyramids,* pp. 113, 132.

1) The mere existence of the Megalithic measures, their accuracy, the inclusion within them of accurate and sophisticated astronomical and physics data — including the velocity of light and the mass of the Earth — point to the existence of a surviving elite from a civilization at least as advanced as our own.

   This observation requires some commentary. In *Babylon's Banksters: The Alchemy of Deep Physics, High Finance, and Ancient Religion,* I noted that the bullion exchange policies of the ancient Occident — Babylonia, Rome, and so on — were oddly coordinated with the bullion exchange policies of the ancient Orient — India and China. This, I noted, would have led inevitably to the rise of an international mercantile and bullion-trading class. I then asked two questions: "Is it possible that, rather than such a class having emerged as a consequence of such governmental policies and trade, that the converse is true? Is it possible that there existed such a class of "international bullion brokers" who *created* these policies in various parts of the world, policies which would enhance their own power and wealth? If so, then how did they achieve and orchestrate this?"[78] The answer may now be given, for in view of the fact that such an elite would have required accurate units of measure to even be able to *conduct* trade, the early emergence of the use of such measures long before the rise of classical civilizations compels to the conclusion that in fact that elite *did* pre-exist those policies and the cultures that created them *by several millennia.* Moreover, as we shall see in the last chapter of this book, there is a significant piece of evidence that trade itself existed long before the establishment of these units of measure in Neolithic times.
2) The existence of an artificial moon — Iapetus — parked in a very "Sumerian" orbit around Saturn suggests strongly that the level of scientific and technological achievement of that civilization was much higher than our own;
3) The elite or elites that survived that ancient cosmic destruction eventually engaged in the "re-measuring of the structure of the Deep," a step necessarily disclosing two possible agendas, and thereby, two possibly competing elites:
   a) The establishment of accurate units of measure of distance,

78 Farrell, *Babylon's Banksters: The Alchemy of Deep Physics, High Finance, and Ancient Religion,* p. 162.

volume, and weight could only be accomplished by the constancy of astronomical observation, and was a necessary step to the establishment of trade sufficient to sustain advanced civilizations and eventually to restore a global civilization. The restoration of such a civilization was necessary if those elites were ever to return to a similar apogee of social and scientific development. If this agenda represents that of one elite, then one may conclude that the agenda is relatively benign, being simply one of ushering humanity to, or *back* to, a similar pitch of social, spiritual, and scientific development;

b) The accurate measuring of "the Deep" was also a necessary step to the restoration of the sophisticated science and technology with which that ancient "Cosmic War" was fought, and possibly also to the use of any technologies that survived from that war. If this agenda represents the agenda of a different elite, then one may conclude that this agenda is malign and for the sole purpose of enhancing the power of the elite pursuing it by reacquiring the technology of hegemony. Needless to say, this activity would also require the reestablishment of trade and civilization, so it becomes difficult to disentangle these two elites and their agendas — if indeed there *are* two — on the basis of the measuring activity alone.

Thus far, of course, we have evidence only of the measuring activity, and therefore of the first or benign elite's agenda. It is when we turn to consider the other major preoccupation of that "post-Cosmic War" elite — religion — that the plot, and the agendas, begin to thicken considerably.

Three

THE TECHNOLOGIES OF SPECIAL REVELATION:

MIND MANIPULATORS, TORSION TEMPLES, AND RELIGION REVEALERS

"...In 1973, Dr. Joseph Sharp, of the Walter Reed Army Institute of Research, expanded on Frey's work in an experiment where the subject — in this case, Sharp himself — 'heard' and understood spoken words delivered via a pulsed-microwave analog of the speaker's sound vibrations."
—Martin Cannon[1]

"In some patients electrical stimulation of the exposed temporal lobe has produced the perception of music. Occasionally it was a determined tune which could be recognized and hummed by the subject, and in some cases it was as if a radio or record were being played in the operating room. The sound did not seem to be a recollection but resembled an actual experience in which instruments of an orchestra or words of a song were heard."
—Jose M.R. Delgado, M.D.[2]

WHEN ONE INTRODUCES the assumption of a very ancient, Very High Civilization, with a science and technology exceeding that of our own, into the picture of ancient history and the civilizations of Sumer and Egypt, the entire basis of interpreting their monuments and texts changes, for the pos-

1 Martin Cannon, "The Controllers: A New Hypothesis of Alien Abduction," www.constitution.org/abus/controll.htm, p. 16.

2 Jose M.R. Delgado, M.D., *Physical Control of the Mind: Toward a Psychocivilized Society* (New York: Harper Colophon Books, 1971), p. 150.

sibilities of interpretation — and agendas at work behind the scenes — increases exponentially. This is nowhere more true than of religion itself, for if one assumes the existence of such scientific sophistication, then perforce one assumes the existence of what might be termed "the oracular technologies of special revelation," or at least of the possibility of constructing them.

One event in modern times highlights, as does no other, the possibilities of the "oracular technologies of revelation," and of the implications of those technologies for the claims of ancient revelations. That the technologies currently exist to manipulate the mind, the emotions, and to do so at a distance is no longer in doubt. But what most people do not realize is that those technologies have been taken far beyond the induction of *generalized* psychological states through direct physiological stimulation of the human brain. Those technologies are now capable of making people think they are seeing *specific* visions and hearing voices that say *specific* things. They are powerful technologies that, used with or without conjunction with other technologies, could even lead people to believe that they are receiving special revelations and instructions from God Himself.

The ethical problem posed by the existence of such technologies is further compounded when one considers the moral problems they engender, i.e., when one considers what those "special revelations" might be telling people to do, either to themselves, or to others. Additionally, their *possible* existence and use in ancient times poses difficult theological and apologetic problems for religions claiming a basis on the special revelations given to isolated individuals, as recorded in any number of sacred scriptures of various religions.

Outlining such a case, however, is fraught with difficulties, since so few people know of the existence of such technologies, much less of the degree of perfection that they are known to exhibit, much less the actual degree of development which remains cloaked in secrecy. Fewer still are aware of the distinctive datasets that eerily connect the details of claimed ancient special revelations and oracles with the capabilities of those modern technologies. Accordingly, we shall outline them as thoroughly as possible, and then turn to the unusual arguments and data points that connect the modern technologies to the indications of their possible use in ancient times. Again, the reader is reminded of what was said in the preface to this book: these possibilities of interpretation are raised to highlight the looming problems for religious apologetics, and those problems will only grow and increase, not decrease, with time, as the technological sophistication of modern society lays bare ever more and more interpretive possibilities in ancient texts.

The modern event that highlights the dangerous potentialities of such technologies within the context of religion is the siege of the Branch Davidian compound in Waco, Texas, in 1993.

A. David Koresh Hears the Voice of God... or Was it Just Charlton Heston?

David Koresh, leader of the Branch Davidian religious compound in Waco, Texas, was (so the official story goes) a man who claimed a special and unique status as a leader within his community.[3] According to some newspaper and magazine articles he saw himself as a veritable second Messiah, uniquely plugged into the mind and thoughts of God. From such a privileged position within his community, he led his followers to a ruinous conflagration and death... or so the story goes.

But few people now recall that during the numerous interviews of "official spokesmen" on the major networks, those spokesmen made no secret that various forms of psychological warfare were being deployed against the Branch Davidians under siege inside their compound, from everything from loud rock music being played 24 hours a day, to other techniques of "inducement."

One of those techniques was exposed by researcher Jon Ronson in a book that has recently been made into a movie, *The Men Who Stare at Goats,* and it affords our entry into the topic of the "oracular technologies of revelation," or, to put it more succinctly and baldly, mind manipulation. The question that inspired his hunt, and that eventually led him to the Branch Davidians, was innocent enough: "Was there," he asked, "somewhere out there, a paper trail of patents for subliminal sound technologies, or frequency technologies, that simply vanished into the classified world of the United States government?"[4] The answer that he found was *not* so innocent:

> On October 27, 1992, Dr. Oliver Lowery, of Georgia, was the recipient of U.S. Patent #5,159,703. His invention was something he called a Silent Subliminal Presentation System:
>
> "A silent communications system in which non-aural carriers, in the very low or very high audio-frequency range of the adjacent ultrasonic frequency spectrum, are amplitude- or frequency-*modulated with the desired intelligence* and propagated acoustically or vibrationally, *for inducement into the brain, typically through the use of loudspeakers, earphones, or piezoelectric transducers.* The modulated carriers may be transmitted directly in real time or may be conveniently recorded and stored on mechanical, magnetic or optical media for delayed or repeated transmission to the *listener.*"[5]

3 The present author has grave misgivings about many aspects of the "official story."

4 Jon Ronson, *The Men Who Stare at Goats* (New York: Simon and Schuster, 2004), p. 179.

5 Ibid., pp. 179–180, citing U.S. Patent Number 5,159,702, my emphasis.

Note the euphemistic phrase "modulated with the desired intelligence" and the equally euphemistic reference to the "listener" of these amplitude- and frequency-modulated "subliminal presentations," for amplitude and frequency modulation are, of course, the two types of modulation in use in AM and FM radios, and modulation *itself* is simply the fancy term for the *information being encoded into the radio waves,* that is to say, for the information you hear when you turn on your radio. Thus, the phrase "modulated with the desired intelligence" really simply means "modulated with the desired *information*" that is being "subliminally presented" to the "listener."

Just what was the "desired intelligence" or "information" to be subliminally presented to the "listener"? According to Ronson,

> The following emotional states could, according to Lowery, be induced by his invention:
>
> *Positive emotions:* contentment, duty, faith, friendship, hope, innocence, joy, love, pride, respect, self-love, and worship.[6]

This is an interesting list in and of itself, given our hypothesis that such technologies might possibly underlie certain special revelations in ancient times. But what of the "negative" emotions? Lowery's list here is as equally disturbing as the so-called "positive" ones:

> *Negative emotions:* anger, anguish, anxiety, contempt, despair, dread, embarrassment, envy, fear, frustration, grief, guilt, hate, indifference, indignation, jealousy, pity, rage, regret, remorse, resentment, sadness, shame, spite, terror, and vanity.[7]

But it did not end there, as Ronson soon discovered.

A mere four years later, on December 13, 1996 to be precise, Ronson discovered that Lowery's company, called Silent Sounds Inc., had posted a message on its website that made for some very disturbing reading:

> "All schematics have [now] been classified by the U.S. Government and we are not allowed to reveal the exact details... we make tapes and CDs for the German Government, even the former Soviet Union countries! All with the permission of the U.S. State Department, of course.... The system was used throughout Operation Desert Storm

6 Ronson, op. cit., p. 180, emphasis in the original.

7 Ibid., emphasis in the original.

> (Iraq) quite successfully."[8]

Apparently Dr. Lowery's invention had found some rather interesting "buyers" — the United States, Germany, and "former Soviet countries" — as well as some rather unique "listeners," presumably Iraqi civilians and soldiers during the First Gulf War.

But what did all of this have to do with David Koresh and the "evil" Branch Davidians under siege in Waco, Texas, ready to break out and storm the M1A1 Abrams tanks with their assault rifles?

Ronson dug further with Dr. Lowery, and soon Lowery mentioned the name of a Russian researcher who had invented a very similar technology, Dr. Igor Smirnov.

> I looked Dr. Smirnov up. I found him in Moscow. I corresponded with his office, and his assistant (Dr. Smirnov speaks little English) told me the following curious story.
>
> It is a story the FBI has never denied.
>
> Igor Smirnov was not prospering in the post-Cold War Moscow of 1993. His finances were so bleak that when the Russian mafia turned up at his laboratory one evening, pressed the bell marked, somewhat ominously, "Institute for Psycho-Correction," and told Igor they'd pay him handsomely if he could subliminally influence certain unwilling businessmen to sign certain contracts, he almost accepted their offer. But in the end it seemed just too frightening and unethical and he turned the gangsters down. His regular clients — the schizophrenics and the drug addicts — may have been poor payers but at least they weren't the mafia.
>
> Igor's day-to-day work in the early 1990s was something like this: A heroin addict would turn up at his lab very upset because he was a father-to-be but try as he might he cared more about the heroin than his unborn child. So he'd lie on a bed, and Igor would blast him with subliminal messages. *He'd flash them on a screen in front of the addict's eyes and blast them through earphones, disguised by white noise,* and the messages would say "Be a good father. Fatherhood is more important than heroin." And so on.
>
> This was a man once fêted by the Soviet government, which — ten years earlier — had instructed him to blast his silent messages at Red Army troops on their way to Afghanistan. Those messages said,

8 Ibid., citing Lowery's website statement, p. 180.

"Do not get drunk before battle."[9]

Then, however, the story began to get very interesting, and with it, the implications of such technologies for claims of revelation multiply like rabbits:

> But the glory days were long gone by March 1993 — the month Igor Smirnov received a telephone call, out of the blue, from the FBI. Could he fly to Arlington, Virginia, right away? Igor Smirnov was intrigued, and quite amazed, and he got on a plane.
>
> The U.S. intelligence community had been spying on Igor Smirnov for years. *It seemed he'd succeeded in creating a system of influencing people from afar — putting voices into their heads,* remotely altering their outlook on life — perhaps without the subjects even knowing it was being done to them.... *The question was: Could Igor do it to David Koresh? Could he put the voice of God into David Koresh's head?*[10]

At this juncture, it is worth pausing to take some stock of the situation, and at some crucial techniques used by Dr. Smirnov.

First, it is to be noted that Dr. Smirnov's technology was taken very seriously, not only by the former Soviet government, but by the American FBI. Secondly, it is also to be noted that the FBI was *not* hesitant to deploy such technology on American citizens. Thirdly, and finally, it is also to be noted that the FBI had a *specific* use in mind for that technology in Koresh's case: it wanted Smirnov to convince Koresh that he was receiving another revelation, hearing the voice of God Himself, while in reality, the voice being heard was only the hidden agenda of the FBI. The pattern — that of an elite in possession of a technology which it is using for a hidden agenda, hiding it behind the "voice of God" — has its own obvious and disturbing implications for the hypothesis of the possible existence and use of such technologies in ancient times.

But one should also take note of two basic facts about the technology — and the technique — used by Dr. Smirnov, for both *optical* and *aural* techniques were used simultaneously and in conjunction with each other. The BBC even ran a short report on two unnamed Russian "psychologists" that were brought in by the FBI. During the report, the Russians' equipment was shown, which included flashing lights on television, as the reporter stated that various key words were beamed in white noise to the "listener." We may refer

9 Ronson, op. cit., pp. 184–185, emphasis added.

10 Ronson, op. cit., p. 185, emphasis added.

to these two significant data points and techniques as the "flashing lights" and "strange sounds" techniques.[11]

In any case, what came of the FBI's attempt to recruit Dr. Smirnov and his technology for use during the Waco massacre?

> The FBI flew Dr. Smirnov from Moscow to Arlington, Virginia, where he found himself in a conference room with representatives of the FBI, the CIA, the Defense Intelligence Agency, and the Advanced Research Projects Agency.
>
> The idea, the agents explained, was to use the telephone lines. *The FBI negotiators would bargain with Koresh as usual but, underneath, the silent voice of God would tell Koresh whatever the FBI wanted God to say.*
>
> *Dr. Smirnov said this was possible.*
>
> But then bureaucracy crept into the negotiations. *An FBI agent said he was concerned that the endeavor might somehow lead to the Branch Davidians committing mass suicide. Would Dr. Smirnov sign something to the effect that if they did kill themselves as a result of the voice of God being subliminally implanted in their heads, he would take responsibility?*
>
> Dr. Smirnov said he wouldn't sign something like that.
>
> And so the meeting broke up.
>
> An agent told Dr. Smirnov it was a shame it didn't work out. He said they had already co-opted someone to play the voice of God.
>
> Had Dr. Smirnov's technology been put into practice at Waco, the agent said, God would have been played by Charlton Heston.[12]

If true — and there is no reason to disbelieve Ronson's research — then this account sheds much light not only into the disturbing possibilities of mind manipulation technology, but also upon the agendas of the elites who would use it, for note the subtle suggestion is that the FBI was perhaps contemplating putting the "suggestion" into Koresh's mind to have everyone inside his compound commit mass suicide à la Jim Jones and the Peoples Temple cult. Had that happened, it would not have been the first time that the voice of "God" has urged wholesale slaughter through a spokesman who has received a "revelation." But more of that later.

---

11 The official U.S. Treasury report on the Waco massacre admits, on p. 17, that it used "loud Tibetan chants on the loudspeakers" and "floodlights" as part of its psychological warfare against the Branch Davidians.

12 Ronson, op. cit., pp. 193–194, emphasis added.

B. The Mind Manipulators: Mind Manipulation Technologies

But was Dr. Smirnov in fact correct? *Could* technology actually be used to induce people to hear the "voice of God" and, by implication, see Him as well? Could those technologies be used, additionally, to induce emotional states of willing obedience and to suppress the individual conscience and will, if not to alter it completely? It is here that the story and its implications become even more — to coin a pun — mind-boggling.

1. The Alien Abduction Scenario and the Moral Disconnect

Martin Cannon is a researcher who has pulled together the most salient articles and books on the subject of mind control and compiled them into an intriguing examination of the alleged phenomenon of "alien abduction" in an Internet article entitled "The Controllers: A New Hypothesis of Alien Abduction."[13] However, the philosophical implications of his observations certainly spill far beyond the containers of "alien abduction" and raise serious questions for the whole prospect of religious revelations as a technologically manipulated phenomenon. Cannon observes that within the field of ufology

> the term "abduction" has come to refer to an infinitely confounding experience, or matrix of experiences, shared by a dizzying number of individuals, who claim that travelers from the stars have scooped them out of their beds, or snatched them from their cars, and subjected them to interrogations, quasi-medical examinations, and "instruction" periods. Usually, these sessions are said to occur within alien spacecraft; frequently, the stories include terrifying details reminiscent of the tortures inflicted in Germany's death camps. The abductees often (though not always) lose all memory of these events; they find themselves back in their cars or beds, unable to account for hours of "missing time."[14]

But the oddest fact that seems to be a common feature of all these stories is that "many abductees, for all their vividly recollected agonies, claim to love their alien tormentors."[15] In other words, there is a "moral disconnect" between the actual experience itself, and the feelings its victims experience or recall toward their perpetrators, reactions that, under ordinary circumstances,

13 I am indebted to my friend Jay Weidner for bringing this important article to my attention.

14 Martin Cannon, *The Controllers: A New Hypothesis of Alien Abduction,* www.constitution.org/abus/controll.htm., p. 1.

15 Ibid.

normal persons would not feel. As will be seen subsequently, this "moral disconnect" is a common feature within "post-revelation" occurrences within religion, and even then, when "ordinary" emotions appear to surface within the religious "experiencer," these are quickly dealt with by a variety of techniques.

2. The False Dialectic, Occam's Razor, and Their Implications

Cannon maintains that a false dialectic of interpretive possibilities has been deliberately created around the abduction phenomenon, a dialectic created precisely in order to deflect attention from a possible interpretation that explains the phenomenon and its real origins. That dialectic is between the "extraterrestrial believers" on the one hand, and the "abduction skeptics" who would maintain that nothing other than particularly vivid nightmares are occurring on the other: The myth of the UFO has provided an effective cover story for an entirely different sort of mystery. Remove yourself from the Believer/Skeptic dialectic, and you will see the third alternative."[16] That third alternative is to consider that (1) the phenomenon itself is very real, but (2) what produced it was a terrestrial technology and agenda, not an extraterrestrial one:

> I posit that the abductees *have* been abducted. Yet they are also spewing fantasy — or, more precisely, they have been given a set of lies to repeat and believe. If my hypothesis proves true, then we must accept the following: the kidnapping is real. The fear is real. The pain is real. The instruction is real. But the little grey men from Zeta Reticuli are *not* real; they are constructs, Halloween masks meant to disguise the real faces of the controllers. The abductors may not be visitors from Beyond; rather, they may be a symptom of the carcinoma which blackens our body politic.[17]

This "believer-skeptic" dialectic is *strongly* suggestive of a similar method that I believe was put into place to control the interpretive possibilities of another famous UFO-related event: the Roswell incident of July 1947.

There, once again, the interpretive possibilities were quickly promulgated by the United States Army Air Force within mere hours of each other, for it maintained that what had crashed, and what it had recovered, were either (1) a flying saucer (with all its "extraterrestrial" implications), or (2) a mere weather

16 Cannon, *The Controllers,* p. 3.

17 Ibid. Another alternative of course — since we are exploring the technological possibilities — is that the little grey men are also real, but they're just not from "out there" but the bizarre products of some genetic technology "down here."

balloon. These two poles established what I call "The Roswell Dialectic."[18] What both dialectics are designed to do is to conceal the possibility of very terrestrial technological explanations for the two events, and to conceal the possible agendas those technologies might suggest.

This dialectic immediately reveals serious ramifications for the technological possibilities of explanation of various "special revelations" in religious history, for one need but substitute the word "God" in the "ET" pole of the dialectic to disclose yet another possibility of locking interpretations into two mutually exclusive poles to conceal a possibly hidden third technological alternative: God becomes opposed to "skepticism," that is to say, to atheism and agnosticism with their "purely mundane" trivializations of what ancient texts state about their special "revelations."

Cannon states this difficulty in connection with Occam's razor:

> Certainly, we are not being narrow-minded if we ask researchers to exhaust *all* terrestrial explanations before looking heavenward.
>
> Granted, this particular explanation may, at first, seem as bizarre as the phenomenon itself. But I invite the skeptical reader to examine the work of George Estabrooks, a seminal theorist on the use of hypnosis in warfare, and a veteran of Project MKULTRA.[19] Estabrooks once amused himself during a party by covertly hypnotizing two friends, who were led to believe that the Prime Minister of England had just arrived; Estabrooks' victims spent an hour conversing with, and even serving drinks to, the esteemed visitor. For ufologists, this incident raises an inescapable question: If the Mesmeric arts can successfully evoke a non-existent Prime Minister, why can't a representative from the Pleiades be similarly induced?[20]

And by the same token, if non-existent Prime Ministers and Pleiadeans can be induced, why not Moses' Burning Bush or Mohammed's visions of Gabriel, and so on?

There is yet another thorny implication posed by the existence of such technologies and techniques for religious apologetics, and that is the "disposal problem." During the era of the CIA's initial mind-control experiments, there was a problem: how to "dispose" of the victims of the experiments and thus keep the experiments secret? As outright murder was considered to be morally repugnant, the solution quickly devolved into erasing the memory

18 Q.v. my *Roswell and The Reich: The Nazi Connection* (Adventures Unlimited Press, 2010), pp. 315–318.

19 MKULTRA: one of the CIA's many covert mind-control projects.

20 Cannon, *The Controllers*, p. 4.

of the event, or implanting entirely false contexts within the victims' minds by which to interpret it. Thus a whole new chapter in mind-control experimentation began, namely, how to wipe selective memories and replace them with false ones? And here, an experience of "God" will work just as well as an experience with "ET," raising yet more possibilities.

3. Electronic Methods of Mind Manipulation

So what exactly are the known technologies of mind manipulation? How do they work? And what are their capabilities? We have already, in our examination of their planned and actual use on David Koresh and the Branch Davidians, seen some of them. The various means might conveniently be divided into two categories: (1) "soft" ones, involving *techniques* of manipulating the mind into channels of interpretive possibilities, such as was noted in the case of the dialectical channeling of the interpretations of "alien abductions" and the Roswell incident; and (2) "hard" ones, involving actual *technological* manipulations of the mind by various means. Our concentration here will be upon the technological means, as opposed to the techniques.

In the CIA's own studies of the techniques and technologies of mind manipulation — which we may assume is a fairly representative template of the work conducted in other nations — virtually the entire spectrum was experimented upon, from "soft" techniques such as hypnosis, drugs, the creation and manipulation of religious cults, extra-sensory perception (ESP), sensory deprivation, and conditioning, to "hard" technologies such as the use of microwaves, brain implants, psychosurgery, and all possible combinations of them.[21] These experiments included projects for

> the "erasure of memory, hypnotic resistance to torture, truth serums, post-hypnotic suggestion, rapid induction of hypnosis, electronic stimulation of the brain, non-ionizing radiation, *microwave induction of intracerebral "voices,"* and a host of even more disturbing technologies.[22]

Whatever the "host of even more disturbing technologies" may have been, it is crucial to note here the use of microwaves to induce the actual "hearing" of voices into the human brain. The implication, both for the "ET abduction" scenario and for the technological potentialities of "revelation manipulation" are rather obvious.

---

21 Cannon, *The Controllers,* p. 5.

22 Ibid., p. 3, emphasis added.

a. Electromagnetic Fields, Implants, and Combinational Approaches

The earliest types of electric manipulation of the brain included direct implantation of electronic components in the brain. Noting that abductees "often describe operations in which needles are inserted into the brain" and "more frequently still, they report implantation of foreign objects through the sinus cavities," Cannon observes that as abduction researchers leap to the extraterrestrial conclusion from these bizarre circumstances, they "have failed to familiarize themselves with certain little-heralded advances in terrestrial technology."[23]

> The abductees' implants strongly suggest a technological lineage which can be traced to a device known as a "stimoceiver," invented in the late '50s-early '60s by a neuroscientist named Jose Delgado. The stimoceiver is a minature depth electrode which can receive and transmit electronic signals over FM radio waves. By stimulating a correctly positioned stimoceiver, an outside operator can wield a surprising degree of control over the subject's responses.
>
> The most famous example of the stimoceiver in action occurred in a Madrid bullring. Delgado "wired" the bull before stepping into the ring, entirely unprotected. Furious for gore, the bull charged toward the doctor — then stopped, just before reaching him. The technician-turned-toreador had halted the animal by simply pushing a button on a black box, held in the hand.[24]

This episode, recounted many times in the literature of mind control, palpably demonstrated the potentialities of the technologies, even in the late 1950s, so it is a simple matter to extrapolate what might be done now, decades later, after enough money, research, and manpower, and those willing to throw morality and humanity out the window in order to achieve breakthroughs.

Delgado pressed his research, and was able by 1973 to report that "Radio stimulation of different points in the amygdala and hippocampus... produced a variety of effects, including pleasant sensations, elation, deep, thoughtful concentration, odd feelings, super relaxation, *colored visions*, and other responses."[25] Note that by 1973, direct contact of an implanted device in the brain was no longer necessary; the effects, including "colored visions," could

---

23 Cannon, *The Controllers*, p. 2.

24 Ibid., p. 6.

25 J.M.R. Delgado. "Intracerebral Radio Stimulation and Recording in Completely Free Patients," *Psychotechnology* (Robert L. Schwitzgebel and Ralph K. Schwitzgebel, editors; New York: Holt, Rinehart, and Winston, 1971), p. 195, emphasis added. Cited in Cannon, *The Controllers*, p. 6.

be induced remotely from a distance via electromagnetic targeting of certain regions of the brain.

These types of discoveries soon ushered in a new quest within the technologies of mind manipulation, technologies that could induce *specific* auditory, visual, and emotional effects in an individual or group target.

> According to a [Defense Intelligence Agency] report released under the Freedom of Information Act, microwaves can induce metabolic changes, alter brain functions, and disrupt behavior patterns. [Project Pandora] discovered that pulsed microwaves can create leaks in the blood/brain barrier, induce heart seizures, and create behavioral disorganization. In 1970, a RAND Corporation scientist reported that microwaves could be used to promote insomnia, fatigue, irritability, *memory loss, and hallucinations.*
>
> Perhaps the most significant work in this area has been produced by Dr. W. Ross Adey at the University of Southern California. He determined that behavior and emotional states can be altered without electrodes — simply by placing the subject in an electromagnetic field. By directing a carrier frequency to stimulate the brain and using amplitude modulation to "shape" the wave into a mimicry of a desired EEG frequency, he was able to impose a 4.5 cps theta rhythm on his subjects — a frequency which he previously measured in the hippocampus during avoidance learning.[26]

Note again that all that is needed is to entrain a wave onto a target brain in the desired frequency of a normal electro-encephalogram and this process can be used to inflict memory loss, hallucinations (seeing visions), and even inflict heart attacks from a distance, and leave no readily apparent evidence that this has been done!

b. The Remote Induction of Trances and "Hearing Voices"

But such technologies and techniques were taken *even further* in work done *in 1973!*

> Trance may be remotely induced — but can it be directed? Yes. Recall the intracerebral voices... of Delgado. The same effect can be produced

26 Cannon, *The Controllers,* p. 15, emphasis added, citing R.J. MacGregor, "A Brief Survey of Literature Relating to Influence of Low-Intensity Microwaves on Nervous Function" (Santa Monica: RAND Corporation, 1970).

> by "the wave." Frey demonstrated in the early 1960s that microwaves could produce booming, hissing, buzzing, and other intra-cerebral static (this phenomenon is now called "the Frey effect"); in 1973, Dr. Joseph Sharp, of the Walter Reed Amy Institute of Research, expanded on Frey's work in an experiment where the subject — in this case, Sharp himself — *"heard" and understood spoken words delivered via a pulsed-microwave analog of the speaker's sound vibrations.*
>
> Dr. Robert Becker comments that "Such a device has obvious applications in covert operations designed *to driver a target crazy with 'voices' or deliver undetectable instructions to a programmed assassin." In other words, we now have,* ***at the push of a button,*** *the technology either to inflict an electronic* ***gaslight*** *— or to create a true* ***Manchurian Candidate.*** Indeed, the former capability could effectively disguise the latter. Who will listen to the victims, when electronically induced hallucinations they recount exactly parallel the classical signals of paranoid schizophrenia and/or temporal lobe epilepsy?[27]

In other words, all the technologies are now in place to induce in a human target, via microwave radiation of particular frequencies, *all of the characteristic features of religious revelations: dazzling light displays, unusual sounds, music, and even voices communicating in real words.*

Yet it does not stop there, for these technologies, *combined with the "soft" technique of hypnosis*, produced even more startling results:

> Perhaps the most ominous revelations, however, concern the mysterious work of J.F. Schapitz, who in 1974 filed a plan to explore the interaction of radio frequencies and hypnosis. He proposed the following:
>
> "In this investigation it will be shown that the spoken word of the hypnotist may be conveyed by modulated electro-magnetic energy *directly into the subconscious parts of the human brain* — i.e., without employing any technical devices for receiving or transcoding the messages and without the person exposed to such influence having a chance to control the information input consciously."
>
> He outlined an experiment, innocent in its immediate effects yet chilling in its implications, whereby subjects would be implanted with the subconscious suggestion to leave the lab and buy a particular item; this action would be triggered by a certain cue word or action.

27 Cannon, *The Controllers,* p. 16, italicized emphasis by me, bold and italicized emphasis by Cannon, citing Robert O. Becker, *The Body Electric* (New York: William Morrow, 1985), pp. 318–319.

> Schapitz felt certain that the subjects would rationalize the behavior — in other words, the subject would seize upon any excuse, however thin, to chalk his actions to the working of free will...
>
> Schapitz's work was funded by the Department of Defense. Despite FOIA requests, the results have never been publicly revealed.[28]

In other words, in addition to having all the "technologies of revelation inducement," if Schapitz's experiments were successful — and their continued classification suggests that they were — then those technologies could also be used to induce certain *actions* in *response* to those "revelations." All of these techniques' and technologies' effects, it was further discovered, could be *amplified* by the presence of an implant in the victim.[29]

c. Electronic Dissolution of Memory: Missing Time, and Missing History

As previously noted, one of the major problems that early mind manipulation research encountered was the "disposal" problem, i.e., how would the perpetrators of the experiments keep the experiments secret? One way, of course, was to simply have the participants murdered, but the other way, erasure of certain parts of their memory, was a more efficient "method" and one which, if means for doing so could be developed, would have its own benefits in the arsenal of mind manipulation. The goal was "EDOM," or "Electronic Dissolution of Memory." It is accomplished by nothing more complicated than the "blockage of synaptic transmission in certain areas of the brain,"[30] similar to what happens in an ordinary stroke. The effect, again, can be produced by electromagnetic "jamming" of the signals of neural pathways, with the result that there is an "erasure of memory from consciousness" in "certain areas of the brain."[31] The result is "missing memory," which, if one looks at it a different way, is nothing but the abductees' "missing time" phenomenon, for by erasing memory, one is erasing *time,* and by erasing time — if performed on a large enough group of people, one is perforce erasing *history.*[32]

---

28 Cannon, *The Controllers,* p. 16, citing Robert O. Becker, *The Body Electric,* p. 321, emphasis Cannon's.

29 Ibid., p. 17.

30 Cannon, *The Controllers,* p. 11.

31 Ibid.

32 Cannon notes that NASA was one of the U.S. government agencies involved in the early mind manipulation research (q.v. p. 3). One possible reason for its interest was disclosed by Richard C. Hoagland and Mike Bara in their *New York Times* bestselling book, *Dark Mission: The Secret History of NASA,* for if the Apollo astronauts found and saw evidence of ancient civilization and technology during the Apollo Moon landings, or even brought some of it back to Earth with them, then this would be a closely guarded secret, and the astronauts' selective memories of what they saw would have been "wiped." Hoagland and Bara comment that the use of such mind manipulation technologies may

Without certain memories, a person's — or a *group* of persons' — context for interpreting events and making decisions will inevitably change, so memory erasure can also be viewed as a means of social manipulation for the purpose of producing a certain class of actions. Again, the obvious implications for the putative "technologies of revelation" is rather obvious.

But there is another possibility that such electromagnetic mind manipulation might have, and within the context of the possibility of truly planet-busting weapons, it is a frightening one, for one must ask: would a normal, sane person press a button and blow up a planet, à la George Lucas' celebrated "Death Star" in his epic movie *Star Wars*? Probably not. What is required is no longer a "normal soldier," but an *insane* one:

> There have always been recruits for even the most hazardous duties; what need of hypnosis?
>
> The need, in fact, is absolute.
>
> The modern battlefield has little place for the traditional soldier. Advanced weaponry requires an increasing level of technical sophistication, which in turn requires a cool-headed operator. But the all-too-human combatant — though capable of extraordinary acts of courage under the most stressful conditions imaginable — does not possess inexhaustible reserves of *sang-froid*.... As Richard Gabriel, the excellent historian of the role of psychiatry in warfare, writes:
>
> "Modern warfare has become so lethal and so intense that only the already insane can endure it...."
>
> ....
>
> According to Gabriel, the military intends to meet this challenge by creating "the chemical soldier," a designer-drugged zombie in fighting man's uniform...[33]

But there may be no need for drugs at all, but merely the requisite electromagnetic field or "template" over a given region with a given target group of "soldiers," for we have already seen the ability of this technology to produce emotional states and actions otherwise not normally experienced by an ordinary person, which would include the ability to induce "insane rages" for no reason. Why use drugs, which will leave traces — evidence — in the body, when electromagnetic waves will do the same trick, and produce no detectable traces?

---

account for the peculiar inability of the Apollo astronauts to remember in any details what they actually *saw* on the Moon. See also John Marks, *The Search for the "Manchurian Candidate": The CIA and Mind Control: The Secret History of the Behavioral Sciences* (New York: W.W. Norton and Co., 1979), p. 225.

33 Cannon, *The Controllers*, p. 29, citing Lincoln Lawrence, *Were We Controlled?* (University Books, 1967), p. 38.

d. "Abductions" and "Revelations:" A Common Method

Strangely enough, there is more than just a passing similarity between the "alien abduction" experience and religious revelations, for as Cannon notes, many of those who have been victims of the abduction experience — and let's be honest, they *are* victims — "have been directed to join certain religious/ philosophical sects."[34]

> I strongly urge abduction researchers to examine closely any small "occult" groups an abductee might join. For example, one familiar leader of the UFO fringe — a man well-known for his espousal of the doctrine of "love and light" — is Virgil Armstrong, a close personal friend of General John Singlaub, the notorious Iran-Contra player, who recently headed the neo-fascist World Anti-Communist League. Armstrong, who also happens to be an ex-Green Beret and former CIA operative, figured into my inquiry in an interesting fashion: An abductee of my acquaintance was told — by her "entities," naturally — to seek out this UFO spokesman and join his "sky-watch" activities, which, my source alleges, included a mass channeling session intended to send debilitating "negative" vibrations to Constantine Chernenko, then the leader of the Soviet Union.[35]

With the tenuous connection of all this bizarre and goofy activity to the World Anti-Communist League, headquartered in Taiwan at that time, the historical circle closes on the place that modern mind manipulation investigation began in earnest — Nazi Germany — as Cannon's descriptions of the organization as "Neo-Fascist" is truer than he perhaps guessed, for the organization is now well-known to have been one of the havens of postwar Nazi activity, ostensibly in behest of its American "masters."[36]

Well might Cannon note the connection between mind manipulation, abductions, and cultic activity, for the data of two of them — mind manipulation and abductions — overlap: "If we could chart these phenomena on a Venn diagram, we would see a surprisingly large intersection between the two circles of information."[37] In other words, there are too many overlapping data points for the parallels between them to be entirely coincidental; the mind manipulation thesis, he notes,

---

34 Cannon, *The Controllers,* p. 29.

35 Ibid., p. 37.

36 Q.v. Martin A. Lee, *The Beast Reawakens: Fascism's Resurgence from Hitler's Spymasters to Today's Neo-Nazi Groups and Right-Wing Extremists* (New York: Routledge, 2000), p. 189.

37 Cannon, *The Controllers,* p. 31.

> explains the reports of abductee intracerebral implants (particularly reports involving nosebleeds), unusual scars, "telepathic" communication (i.e., externally induced intracerebral voices) concurrent with or following the abduction encounter, allegations that some abductees hear unusual sound effects (similar to those created by the hemi-synch and cognate devices), haywire electronic devices in abductee homes, personality shifts, "training films," manipulation of religious imagery, and missing time.[38]

This same methodology — comparison of detailed data points between the mind manipulation technologies and the details of "religious revelations" — will inform the final section of this chapter.

4. The Beat Frequency of the Brain

"The brain," notes Cannon, "has a 'beat' of its own."[39] This fact was discovered by the German psychiatrist Hans Berger in 1924. Berger observed

> two distinct frequencies: alpha (8–13 cycles per second), associated with a relaxed, alert state, and beta (14–30 cycles per second), produced during states of agitation and intense mental concentration. Later, other rhythms were noted, which are particularly important for our present purposes: theta (4–7 cycles per second), a hypnogogic state, and delta (.5 to 3.5 cycles per second), generally found in sleeping subjects.[40]

These publicly available facts became the basis for a whole new "industry" beginning in the late 1970s and into the early 1980s, as clever inventors devised machines like the "hemi-synch," a headphone-like device that produced "slightly different frequencies in each ear" which the brain would then calculate the difference between, resulting in a kind of "beat frequency" to which it would then entrain itself. In other words, "the subject's [electro-encephalogram] would slow down or [speed] up to keep pace with its electronic running partner."[41]

This idea of a beat frequency may seem odd or even unimportant, but it is in fact one of the crucial operational principles of so-called "scalar electromagnetics," and a word or two more of explanation is thus worthwhile here. The whole basis of this type of phenomenon is *interferometry* and the establish-

38 Ibid., p. 31.

39 Cannon, *The Controllers,* p. 13.

40 Ibid.

41 Ibid.

ment of a "beat frequency" as a kind of template for action. And the way to establish that "beat frequency template" is precisely to interfere, or blend, two or more signals of *different* frequencies on a region. The difference between all these interfered or blended frequencies then sets up a "standing wave" that is based on the beat frequency. One may think of the simple — though clumsy — analogy of taking a handful of pebbles of different sizes and throwing them into the calm surface of a pond. Each pebble will produce a wave pattern as it strikes the surface, and those patterns will cross and blend, establishing a momentary pattern on the surface that would represent the standing wave of the beat frequency. It is this pattern or template that the brain entrains itself to, and this point will become crucial in later sections of this chapter.

5. The Beat Frequency, and Out-of-the-Body Experiences

Oddly enough, one of the claims made for some of the "brain machines" that became popular during the late 1970s and early 1980s was that they could also "induce 'Out-of-Body Experiences,' in which the percipient mentally 'travels' to another location while his body remains at rest. This rapidly-developing technology has spawned a technological equivalent to the drug culture..."[42] The apologetic implication for the possible existence of such technologies in ancient times is rather obvious, for such technologies could induce people to believe they have visited "heaven," whether "in the body or not," as the apostle Paul said.[43]

6. The Remote Induction of Trances, Emotions, Specific Information, and Remote "Telepathy"

As work progressed on the technologies of mind manipulation, it was soon discovered — via techniques such as the establishment of "beat frequencies" in a target brain or brains — that emotional states, including deep hypnogogic states, could be remotely induced without actual electrodes or implants having to be present in the brain itself. This technique was called RHIC, or "Remote Hypnotic Intracerebral Control."[44] The phenomenon was first reported by L.L. Vasilev of the University of Leningrad in the early 1930s.[45] Thus, by producing beat frequencies in synch with the alpha, beta,

42 Cannon, *The Controllers,* p. 13

43 Q.v. II Corinthians 12:2–3: "I knew a man in Christ above fourteen years ago, (whether in the body, I cannot tell; or whether out of the body, I cannot tell: God knoweth;) such an one caught up to the third heaven." Most interpreters believe that Paul was referring to himself.

44 Ibid., p. 10.

45 Ibid.

theta or delta states of the brain's natural functioning, the brain would naturally entrain itself to that beat frequency, and experience the corresponding emotional state. Thus, a subject could be made open to the possibility of remote *suggestion* by entraining a beat frequency in the theta cycle range and then beaming "voices" and instructions to the subject, since the same technology could be used, as we have seen, to induce a subject to hear actual "voices" in his brain.[46]

There was yet another possibility that emerged from all this as well, and that was remote electromagnetic *telepathy*, or reading a target's mind or emotional states by "deciphering the brain's magnetic waves." The project was under way since 1973 at the Advanced Projects Research Agency (ARPA).[47]

7. Electromagnetic Alteration of DNA

Beyond all these suggestive and disturbing possibilities and implications for the remote electromagnetic manipulation of the mind, however, there is one more, and that is the ability of prolonged exposure to electromagnetic fields to alter the very characteristics of human DNA itself. While not strictly germane to the subject being discussed here, it is worth noting that researcher Paul Brodeur maintains unequivocally that since the 1960s "the government and the military have systematically suppressed information about the genetic effects of microwaves in human beings and covered up a number of potentially embarrassing situations in which such effects have been observed."[48] Among the effects observed were the long-term permanent damage caused by chromosomal breaks, breaks that in turn were induced by prolonged exposure to such fields.[49]

8. Section Summary

At the end of this excursion into modern mind manipulation techniques and technologies, what do we have? A short review reveals the apologetic implications for the accounts alleged of special revelations in ancient times *if* such a technology is assumed to have existed and to have been used in ancient times, for by relatively simple means, the modern techniques and technologies can remotely induce:

---

46 See also Paul Brodeur, *The Zapping of America* (New York: W.W. Norton & Co.: 1977), p. 295.

47 Brodeur, *The Zapping of America,* p. 299.

48 Ibid., p. 134. See also pp. 120–121.

49 Brodeur, *The Zapping of America,* pp. 104–105.

1) Emotional states of all sorts, from euphoria, anger, hypnogogic suggestibility, trance, and so on;
2) Actual perception of "visions" and "voices" with specific information or instruction;
3) Post-hypnotic suggestibility;
4) Out-of-body experiences;
5) Induction of false memories, "missing time" (and thus, missing *history*), and false interpretive contexts.

The only question now is, is there any evidence that such technologies, howsoever "primitive," might have existed in ancient times? And if there existed an "elite" or "elites with agendas" to manipulate mankind into the earliest stages of civilization as was argued in the previous chapter, is there any evidence to suggest that these technologies were employed to manipulate the most socially cohesive force in human history, religion itself?

The answer, sadly, is yes.

C. The Torsion Temples of Antiquity: A Primitive Technology of Communication and Special Revelation

In seeking to answer the question of whether or not such mind manipulation technologies existed and that they were used in ancient times to manipulate religion itself as a socially cohesive force, our attention must be focused, once again, on "texts," with the understanding that, in a certain sense, monuments and structures themselves are "texts" to be understood and properly decoded by the requisite physics and engineering principles. Additionally, of course, we shall have to contend with actual *written* texts and what they portend in this regard.

In a previous book — *Babylon's Banksters: The Alchemy of Deep Physics, High Finance, and Ancient Religion* — I argued for the existence of a globally-extended financial elite that had attached itself to the temples and priesthoods of antiquity for the purpose of installing a financial monopoly over the private issuance of monetized debt, and for the purpose of manipulating the bullion policies of various empires. As a consequence of this conclusion, I also argued that such manipulation would have required — then as now — a secure means of *rapid* communication between vast distances. That, in turn, would have required the existence of a hidden technology, or, at the minimum, a technology *that was not genuinely appreciated as such* by modern interpreters.[50]

50 For the existence of this elite, see my *Babylon's Banksters: The Alchemy of Deep Physics, High*

Fortunately, I did not have to look long nor far to find an engineer who had tackled precisely this problem, and who had made an amazing discovery, namely, *the technology was the ancient temples themselves, and moreover, that technology was a **scalar** electromagnetic technology, in short, a **radio** technology.*[51] The engineer was a German professor of engineering, Prof. Dr. Konstantin Meyl, and his research was nothing less than stunning and sweeping in its implications. How else, he asked, could one make sense of some statements in classical texts that Roman Empire functionaries in distant regions of the Empire would "send to the Emperor" for instructions, and receive an answer at the latest the following night?[52] Meyl's answer is simply that there was a technology in play: religious temples.

Meyl states the case that ancient temples were actually sophisticated radio transmitters by posing the classic "engineer's optimalization problem":

> Let's to some extent proceed from the knowledge of textbook physics currently present in high frequency engineering and give a well-trained engineer the following task, which he should solve systematically and like an engineer. *He should build a transmitter with maximum range at minimum transmitting power,* thus a classic of optimization. *Doing so, the material expenditure doesn't play a role!*
>
> After mature deliberation the engineer will hit upon it that only one solution exists. He decides on a telegraphy transmitter at the long wave end of the shortwave band, at f=3 MHz, which corresponds to a wavelength of l=100m. There less than 1 Watt transmitting power is enough for radio communication once around the earth....
>
> And he optimizes further. Next the engineer remembers that at high frequencies, e.g. for microwave radiators, not cables but wave-guides are used, since these make possible a considerably better degree of effectiveness. In the case of the waveguide the stray fields are reduced by alignment and concentration of the fields in the inside of the conductor. In the case of antennas, however, the fields scatter to the outside and cause considerable stray losses. He draws the conclusion that his transmitter should be built as a tuned cavity and not as an antenna!
>
> As a result the engineer puts a building without windows in the countryside with the enormous dimensions of 50 m length (=l/2)

Finance, and Ancient Religion (Feral House, 2010), pp. 251–264. For the technological considerations, see pp. 245–265.

51 Farrell, *Babylon's Banksters,* p. 253. See also Prof. Dr.-Ing. Konstantin Meyl, *Scalar Waves,* pp. 608, 610.

52 Farrell, *Babylon's Banksters,* pp. 23–24. See also Meyl, *Scalar Waves,* pp. 631, 615.

and 25 m (=l/4) respectively 12.5 mm (=l/8) width. The height he calculates according to the Golden Proportion to increase the scalar wave part. *Those approximately are the dimensions of the Cella without windows of Greek temples.*

For the operation of such a transmitter in antiquity apparently the noise power of the cosmic radiation was sufficient, which arrived at the earth starting from the sun and the planets. By increasing the floor space also the collected field energy and the transmitting power could be increased, so that also from the perspective of the power supply, the temple with the largest possible wavelength at the same time promised the largest transmitting power, so at least in antiquity.

Our engineer further determines that he will switch the carrier frequency on and off at a predetermined clock pulse. Thus he decides for radiotelegraphy. The advantage of this technique is a maximum of the increase of the reception range. For that the signals at the transmitter have to be coded and at the receiver again deciphered. By means of the encryption of the contents these are accessible only to the "insiders," who know the code; *prerequisite for the emerging of hermeticism and eventually a question of power!*[53]

With these insights in hand, Meyl examined various ancient temples, and calculated the resonant frequencies of the structures themselves, under the understanding that the structure itself was the waveguide.

The results — while I did not comment on them in *Babylon's Banksters* — were astounding, and here it *is* necessary to comment on them, because they are directly germane to the thesis being examined here, namely that these temples might also have functioned not only as transmitters and receivers for microwaves, but as mind manipulation *environments.* For the Temple of Zeus at Olympia, for example, Meyl calculated the frequency of the building to be 5 megahertz (MHz).[54] For the temple of Athena, it was 7.5 MHz.[55] For the Temple of Apollo at Corinth, 9 MHz.[56] For the Temple of Venus and Roma, 6.8 MHz.[57]

Why are these results so interesting? Because if one understands the laws of harmonics, they are all harmonics of the alpha and theta wave states of the brain! A "harmonic" is simply a multiple of a number, and thus, the Temple of Zeus, resonant to 5 MHz, is six orders of magnitude, that is, one million

53 Meyl, *Scalar Waves,* emphasis added, p. 613; cited in *Babylon's Banksters,* pp. 259–260.
54 Meyl, *Scalar Waves,* p. 612; q.v. *Babylon's Banksters,* p. 254.
55 Meyl, *Scalar Waves,* p. 614; q.v. *Babylon's Banksters,* p. 255.
56 Meyl, *Scalar Waves,* p. 614; q.v. *Babylon's Banksters,* p. 256.
57 Meyl, *Scalar Waves,* 620; q.v. *Babylon's Banksters,* p. 256.

times more than 5 cycles per second, which is within the theta stage of brain wave activity, that stage associated with a hypnogogic state.[58] For the Temple of Athena, resonant to 7.5 MHz, this would be a resonance at six orders of magnitude greater, exactly *halfway between* the alpha state of the brain, at 8–13 cycles, a state associated with a relaxed and alert state, and the theta state, again from 4–7 cycles per second. For the Temple of Apollo at 9 MHz, again the resonance is to the alpha state of the brain, and for the Temple of Venus and Roma, the theta state!

To put it succinctly, in those cases examined by Meyl, not only are we dealing with microwave transmitters and receivers, but with structures that appear to have been deliberately designed to induce alpha or theta, or both states of mind, states of "relaxed alertness," or a "hypnogogic" state, or a state of mind somewhere between the two! The same technology, based on pulsed radio waves, was also engineered to produce various emotional states for anyone inside the structures!

So, in answer to the question of whether such technology existed, and whether it was ever deployed to manipulate the mental and emotional environment, our answer must be a tentative "yes." But was that technology ever used in more overt ways and displays to manipulate the minds of men? Was it ever used to mold an entire religious consciousness?

To answer that question, we must turn to the texts themselves, and to the provocative analysis of the husband and wife team of British researchers Christian and Barbara Joy O'Brien, and to the very disturbing questions for religious apologetics that they raise.

D. The Religion Revealers: O'Brien's Technological Consideration of the Revelations of YHWH and the Torah

With the existence of such admittedly primitive but nonetheless effective technologies in ancient times, an existence tied quite directly to the temples — for the technology *was* the temples — the speculative possibility arises that famous (or depending on one's lights, infamous) revelations were the products of some sort of technological manipulation by an elite. One need only recall that all that was required to produce "out of the body experiences," in the case of some "mind machines," was an interference of two waves of different frequency to produce beat frequencies in a brain that would produce a requisite emotional-physiological state. In view of "the techniques and technologies of special revelation," these potentialities, while highly speculative to be sure, do

58 See pp. 100–101.

open the doors to apologetical difficulties posed by famous instances of "special revelation," such as Paul's "vision" of Christ on the road to Damascus, or Daniel's and Joseph's roles in the Old Testament interpreting dreams for their masters: were they dreams they themselves induced and subsequently "interpreted"? As noted in my book *Babylon's Banksters,* the presence of a shadowy financial elite within the precincts of the ancient temple, an elite moreover ready and willing to manipulate religion to its own ends, raises the apologetical and interpretive bar substantially, especially if one entertains the possibility that something of a lost technology and technique of mind manipulation were preserved by that elite.

But one need not only turn to *technological* speculations to perceive the enormous difficulties posed by mind manipulation in the context of religion, for as was previously observed, these constitute but one half of the total picture, the other half being the *techniques* of mind manipulation. And here, according to British researchers Christian O'Brien and his wife Barbara Joy, one need go no further than the Yahweh, or Jehovah, of the first five books of the Old Testament to see the use both of technologies and techniques for the total enslavement of a people, and their turning from normal moral constraints to a willingness to commit wholesale slaughter on behalf of their "divine master," Yahweh.

This is not to say that the O'Briens themselves view Yahweh's technological displays and his behavior toward the Israelites as a manifestation of such technologies and techniques. In fact, they nowhere mention "mind control" or "mind manipulation" in connection with their examination of Yahweh's character and behavior. Nonetheless, it is a disturbing possibility fraught with apologetical implications that clearly emerges from their examination, so it will be reviewed carefully here.

Before reviewing the O'Briens' examination of Yahweh's character and behavior, however, it is worth pausing to take note of the assumptions and the basic methodology that form the context of their interpretation. There are four basic assumptions and methods driving their interpretation of the Yahweh of the Torah, and particularly, of the book of Exodus:

1) There is an ancient elite manipulating the events of human history and religion operating behind the scenes. This elite is connected to and composed of members of the "Anunnaki," a race of "shining ones" whom the O'Briens view as the genetic "cousins" of modern-day *Homo sapiens sapiens,* and indeed, whom they view as having genetically manipulated the latter species into existence. *Some* of that elite rebelled and broke ranks with the rest, interfering in human affairs in a proscribed fashion. For the

O'Briens, the Yahweh of the first five books of the Old Testament is precisely one such individual rebelling, fallen, "shining one."

2) The Shining Ones, representing an advanced, genetically related race to mankind, also possessed an advanced technology which, in Yahweh's case, they believe was used to cow and coerce the Israelites into unhesitating, unquestioning obedience to Yahweh and his plans for conquest.
3) In their examination of the character, behavior, and actions of Yahweh in the Torah, the O'Briens subject him to an unhesitating and thoroughly critical examination based on normal standards of human decency and morality, and find him thoroughly wanting in compassion, love, and exhibiting classical symptoms of paranoia, schizophrenia, and a genocidal blood and war lust comparable to a Stalin or a Hitler.
4) In their examination, the O'Briens make it clear that they believe the name "Yahweh" itself might be a *titular* usage, similar to "the Lord Mayor of London," a usage that is not therefore a proper name, but the name of a *position* within a hierarchy that *different occupants fill at different times*, thus accounting — in the O'Briens' minds — for the vast changes of character and behavior from the Yahweh of the Torah to the Yahweh of, for example, prophets such as Isaiah. It is best to cite their own words here:

> This Yahweh was only one of the leaders of the Shining Ones. The fact that he was always defined as 'Leader' — whether he was taking part in the establishment of the Garden in Eden; pretending to make Abraham sacrifice his eldest son on a woodpile; handing a Code of Laws to Hammurabi; militarily training the Israelites for the conquest of Canaan; or sacrificing the Jewish remnants to the cruel ravages of Babylon — has led to much confusion.
>
> The participant in all these activities was not a single entity.[59]

Our review of the O'Briens' examination of Yahweh will concentrate on these four areas. In order to demonstrate the close comparison of some of Yahweh's behavior toward the Israelites and that of various mind manipulation techniques, that review will of necessity have to be as comprehensive as possible.

59 Christian and Barbara Joy O'Brien, *The Genius of the Few: The Story of Those Who Founded the Garden in Eden* (Dianthus Publishing Limited, 1999), p. 173.

1. Yahweh as One of the Shining Ones

As a representative of the "Shining Ones" whom the O'Briens believe to have been a genetically related race to mankind, and indeed, its ultimate genetic engineers, the assumption that Yahweh is a representative of that race — and therefore *not* God or "a god" in *any* sense — is therefore a study that would yield considerable information about that race. As they put it:

> From a detailed study of this Yahweh, we ought to be able to formulate our ideas on the true characteristics of the Shining Ones — their physical attributes, their temperaments, their technological attainments, and the nature of their relationships with the humans with whom they lived and worked — in much greater depth than we have been able to achieve hitherto.[60]

That, however, did not come without a caveat, for the character of Yahweh — if indeed he was one of the Shining Ones to begin with — was hardly typical of them as a group:

> And yet, we have an uneasy feeling about this Yahweh, and must enter a caveat — he may not have been typical of his race. He was bellicose and vindictive, and appears to have been determined to conquer the peoples of the Middle East by force.[61]

This reading of Yahweh as a vindictive, bellicose conqueror is the characteristic that most identifies him as one of the *rebelling* Shining Ones, a characteristic that will be examined more closely in the second part of this book.

So odd, so *unusual* is Yahweh's behavior when compared to the rest of the Shining Ones that the O'Briens question whether he — and they — are in fact genetically related to humanity at all. "This study of Yahweh may help us to solve the perplexing and disturbing question of whether" he even "belonged to our genus — perhaps *Homo sapiens sapiens sapiens* — or some other genus with which we are not familiar."[62]

2. A Genetic Agenda?

For the O'Briens, the answer to that question is a tentative "yes, the character of Yahweh belongs to a related genus," for as they point out, Yahweh

60 Christian and Barbara Joy O'Brien, *The Genius of the Few,* p. 175.

61 Ibid.

62 Christian and Barbara Joy O'Brien, *The Genius of the Few,* p. 175.

appears to have some sort of hidden *genetic agenda* with respect to the ancient Hebrews, an agenda that is never quite fully revealed, nor one that is ever quite fully concealed either:

> ...before we discuss the enigma [of Yahweh] proper, it must be stressed that the building of the Hebrew nation was achieved by a deliberate process of selection which stretched back to Noah. His family was selected for survival after the Flood; and Terah's family was selected from the Semitic peoples of Ur; and out of it Abraham was chosen, over his brothers, for reasons that we can only guess at. Possibly they were genetic; but another aspect may have been that Sarah, Abraham's wife, was barren — and this may have given the opportunity for another remarkable genetic opportunity.[63]

Later on, of course, another selection was made with Jacob over Esau, this time, note the O'Briens, "by a somewhat underhanded manoeuvre,"[64] referring to Jacob playing a trick on his father Isaac, and receiving Esau's blessing in his place.[65]

All this, to the O'Briens, is suggestive of a hidden genetic agenda:

> The purity of the strain, which Yahweh required, continued to be carefully monitored. He was determined that the line through the Patriarchs from Abraham should continue as he wished it — and he was looking quite a long way into the future. Five hundred years were to pass before he was satisfied that he had achieved the Nation that he wanted.[66]

The implication — even though they do not spell it out in so many words — is clear, for if Yahweh was intent upon conquest, as they believe, then one possible reason for the constant genetic manipulation might be to create a leader-caste with the requisite warrior characteristics. That there are other possibilities for this hidden genetic agenda will be explored in the second part of this book.

3. Yahweh's "Pillar of Fire:" The Technological Component

Central to their thesis that Yahweh, far from being "God" or "a god," is a

63 Ibid., pp. 176–177.

64 Ibid., p. 178.

65 Genesis 27:1–46.

66 Christian and Barbara Joy O'Brien, *The Genius of the Few,* p. 178.

real physical, humanoid creature is the pillar of fire, the technology by which and from which he leads, and intimidates, the Hebrews.

> We are told that he led them from inside an airborne 'pillar of cloud' by day, which became a 'pillar of fire' by night. And here it must be stressed that, after dark, 'fire' and 'light' became synonymous terms because, as we emphasized in the case of Enoch, there was no other light than that made by fire which the chronicler could conceive as illumination, either for the interior of a house, after dark, or for the interior of an airborne object. It is also recorded that the pillar of fire lighted their path by night, which suggests that it was a large, very bright object that spread its light over an extensive area.
>
> (Ex. 13:21) "The Lord went before them in a pillar of cloud by day to guide them along the way and a pillar of fire by night, *that they might travel day and night.*"
>
> At this point it is essential to clarify the issue of the 'pillar' because it is a focal point in any attempt to access the level of technology available to the Shining Ones. We must also consider another quotation, slightly out of chronological order:
>
> (Ex. 14:24) "At the morning watch, the Lord looked down upon the Egyptian army from a pillar of fire and cloud, and threw the Egyptian army into a panic."
>
> There is a gratifying consistency about this short passage because, in the half-light of the morning watch, the airborne object might be expected to show both its daylight form and the light from its source of illumination.
>
> It is essential to this study that we, also, should be consistent by treating the evidence provided by the various chroniclers realistically. If they are lying, or indulging in fantasy, absurdities would very soon show up. But neither the Kharsag epics, nor Enoch, nor the Biblical accounts, nor the Atra-Hasis, has dealt in absurdities — only in apparent anachronisms.
>
> But since all four have consistently described to us a level of technological achievement for the Shining Ones, equal to, or in advance of, our own — we should now accept that they were an advanced race, and cease to be surprised at the wonders which they display.
>
> In the case of the 'pillar of cloud,' the chronicler, who was possibly Moses, was describing a phenomenon outside his normal range of experience, just as Enoch was doing at the House of the Most High,

> but our experience is not so restricted, and we should have a better chance of identifying the object than Moses had.[67]

Note the statement from Exodus italicized in the previous quotation, that Yahweh guided the Hebrews "that they might travel day and night," for this will become a clue, as we shall see, nor only to a technology, but to the *techniques* in play.

4. The Threats Begin: The Carrot, the Stick, and Other Techniques

Having thus interpreted the "pillar of cloud and fire" as a *technology*, the way is now open for the O'Briens to follow up on all the implications that such an interpretation poses, and the first of these is the meeting of Yahweh and the Hebrews at Mount Horeb:

> The people were now given three days to rest, to wash their clothes — in preparation for the awful prospect of meeting Yahweh. Until then, they had had to rely on second- and third-hand reports of this Presence through Moses and their Elders, although the sight of the aerial craft[68] must have been a continual reminder that they were under the direction of an unusual form of leadership.
>
> The setting for this dramatic meeting, and the stage-management of it, must have been magnificent. Well-defined bounds had been set around the base of the mountain which no one — man or beast — was to overstep on pain of instant death. The threats had started!
>
> (Ex. 19:12–13) "Whoever touches the mountain shall be put to death; no hand shall touch him, but he shall either be stoned, or pierced through; beast or man, he shall not live."
>
> At that time Moses was unfamiliar with Yahweh's long-range methods of killing, and could only think in terms of stones, spears or arrows — he was to learn differently.
>
> At Horeb the Israelites had their first taste of Yahweh's powers, the strictness of his orders, and of the penalties for disobedience. The mountain had probably been occupied by Yahweh as a base for a considerable time before the Israelites arrived... As a modern electric fence is used to control cattle, so it might have had its own physical protection against intruders — perhaps a lethal barrier. We incline to

67 Christian and Barbara Joy O'Brien, *The Genius of the Few*, pp. 182–183, emphasis added.

68 i.e., the pillar of cloud and fire.

> this view because of the reference to 'beast or man.' Such stringent regulations would have been unnecessary for the stray goat or dog unless there was physical danger for them on the mountain.
>
> We begin to understand that this was no mountain Eden with the Lord of Spirits strolling amongst his people; this was not Enlil walking through his plantations while men followed his benign instructions. *Here at Horeb, Man was under a very different regime.*[69]

Interestingly enough, during the era of *modern* mind manipulation research, research was also undertaken in forms of crowd control that involved establishing invisible electromagnetic barriers that could — without the aid of electric fences — inflict terrible pain or potentially even death on those attempting to cross or breach it:

> There is, however, a small portion of [a] (Defense Intelligence Agency) report which remains classified. It clearly indicates that efforts to develop microwave radiation as an antipersonnel weapon have been underway in the United States for some years. Take, for example, the following paragraph:
>
> A study published in 1972 by the U.S. Army Mobility Equipment Research and Development Center, titled "Analysis of Microwaves for Barrier Warfare," examines the plausibility of using radio-frequency energy in barrier-counterbarrier warfare. *It discusses both anti-personnel and anti-materiel effects for lethal and non-lethal applications for meeting the barrier requirements* or delay, immobilization, and increased target exposure. The report concludes that:
>
> a. It is possible to field a truck-portable microwave barrier system that will completely immobilize personnel in the open with present-day technology and equipment.
> b. There is a strong potential for a microwave system that would be capable of delaying or immobilizing personnel in vehicles.
> c. With present technology no method could be identified for a microwave system to destroy the type of armored materiel common to tanks.[70]

Note that such a barrier would have potentially *lethal* effects — i.e., be virtually an impermeable invisible barrier to organic life attempting to cross it.

69 Christian and Barbara Joy O'Brien, *The Genius of the Few,* p. 187, emphasis added.

70 Q.v. Paul Brodeur, *The Zapping of America,* p. 296, emphasis added.

It is this "both man and beast" warning that, for the O'Briens, rightly signals the possibility of a technology in play, for such a technology, then as now, would be unlikely to be able to discriminate between the two. Indeed, it is this one fact that most strongly suggests a technology is in play at Horeb, beyond the "normal" pillar of cloud and fire.

It is at Horeb that Yahweh's character and behavior — for the O'Briens — shows itself for what it really is, and it is hardly that of even a relatively enlightened human being, much less a "god." For example, there is the Covenant itself:

> The Covenant between Yahweh and his People was struck at Horeb, and the fact is recorded in Exodus — but we have to wait until the Book of Leviticus before finding out the terms of this Bargain. We shall, however, record its terms here because, without an understanding of what had been agreed, it is impossible to appreciate the vicissitudes that Israel underwent in the Wilderness through Yahweh's interpretation of its clauses.[71]

The O'Briens quip that "Yahweh's verbal interpretation of how he intended to keep his side of the bargain should have given Moses many sleepless nights."[72]

Then come the "Blessings":

> (Lev 26:3–13): "If you follow my laws and faithfully observe my commandments, I will grant you rains in their season so that the earth shall yield its produce and the trees of the field their fruit. Your threshing shall overtake the vintage, and your vintage shall overtake the sowing; you shall eat your fill of bread and dwell securely in your land.
>
> I will grant peace in the land, and you shall lie down untroubled by anyone; I will give the land respite from vicious beasts, and no sword shall cross your land. You shall give chase to your enemies, and they shall fall before you by the sword. Five of you shall give chase to a hundred, and a hundred of you shall give chase to ten thousand; your enemies shall fall before you by the sword.
>
> I will look with favour upon you, and make you fertile and multiply; and I will maintain my covenant with you. You shall eat grain long stored, and you shall have to clear out the old to make room for the new.

71 Christian and Barbara Joy O'Brien, *The Genius of the Few,* pp. 188–189.

72 Ibid., p. 189.

> I will establish my abode in your midst, and I will not spurn you. I will be ever present in your midst; I will be your God, and you shall be my people. I the Lord your God who brought you out from the land of the Egyptians to be their slaves no more, who broke the bars of your yoke and made you walk erect."[73]

Of course, all these blessings can come quite naturally from God. But — and this is the problem posed by the assumption of sophisticated technology — they can come equally from a technology capable of weather manipulation, and notably, such technological sophistication would also be a "multiplier" effect in military terms, easily capable of offsetting overwhelming numerical superiority of an enemy. The O'Briens' point once again, while subtle, is nonetheless clear: such statements are not conclusively indicative of anything divine at all, for there are technological possibilities of interpretation.

Again, the same two possibilities exist, and while reading the following passage, the reader is reminded to read it from the standpoint of someone who is in possession of such a technology dictating terms to someone who is *not:*

> (Lev. 26:14–33): But if you do not obey me and do not observe all these commandments, if you reject my laws and spurn my norms, so that you do not observe all my commandments and you break my covenant, I in turn will do this to you: I will wreak misery upon you — consumption and fever, which cause the eyes to pine and the body to languish; you shall sow your seed to no purpose, for your enemies shall eat it. I will set my face against you; you shall be routed by your enemies, and your foes shall dominate you. *You shall flee though none pursues.*

Let us pause and look at that last statement more closely: "You shall flee though none pursues." Such a statement, given the technological perspective of their overall interpretation, could be very indicative of a mind manipulation technology. To continue:

> And if for all that, you do not obey me, I will go on disciplining you sevenfold for your sins, and I will break your proud glory. I will make your skies like iron and your earth like copper, so that your strength shall be spent to no purpose. Your land shall not yield its produce, nor shall the trees of it yield their fruit.

73 Cited in Christian and Barbara Joy O'Brien, *The Genius of the Few,* pp. 189–190.

This, of course, sounds suspiciously familiar, for the "starvation" tactic was tried by the gods of the Sumerian pantheon long before, to bring down human population![74] Yahweh rages on:

> And if you remain hostile toward me and refuse to obey me, I will go on smiting you sevenfold for your sins. I will loose wild beasts against you, and they shall bereave you of your children and wipe out your cattle. They shall decimate you, and your roads shall be deserted.
>
> And if these things fail to discipline you for me, and you remain hostile to me, I too will remain hostile to you: I in turn will smite you sevenfold for your sins. I will bring a sword against you to wreak vengeance for the Covenant; and if you withdraw into your cities, I will send pestilence among you, and you shall be delivered into enemy hands. When I break your staff of bread, ten women shall bake your bread in a single oven; they shall dole out your bread by weight, and though you eat it, you shall not be satisfied.
>
> But if despite this, you disobey me and remain hostile to me, I will act against you in wrathful hostility...

Say what? The previous list was *not* wrathful hostility?

> ...I, for my part, will discipline you sevenfold for your sins. You shall eat the flesh of your sons and the flesh of your daughters. I will destroy your cult places and cut down your incense stands, and I will heap your carcasses on your lifeless fetishes.
>
> I will spurn you. I will lay your cities in ruin and make your sanctuaries desolate, and I will not savour your pleasing odours. I will make the land desolate to you so that your enemies who settle in it will be appalled by it. And you will scatter among the nations, and I will unsheath the sword against you. Your land shall become a desolation and your cities a ruin.[75]

If Yahweh was one of the "Shining Ones" as spoken of in some ancient cuneiform texts as the O'Briens believe, then at least his rages and threats are true to form, for in the Atra-Hasis, the Mesopotamian Flood epic, as the gods are trying to starve humanity into non-existence, one may read that "They served up a daughter for a meal, served up a son for food."[76]

---

74 See my *The Cosmic War: Interplanetary Warfare, Modern Physics, and Ancient Texts,* p. 146.

75 Cited in Christian and Barbara Joy O'Brien, *The Genius of the Few,* pp. 190–191.

76 Farrell, *The Cosmic War,* p. 146.

For the O'Briens the covenant becomes — just as it does for many devout Jews and Christians — a revelation of Yahweh's character. The only problem is, what they see revealed there is hardly "God" nor even "a god," but something far worse:

> We find this Covenant a most disturbing document on four counts:
>
> 1. This was not a freely negotiated agreement between two parties in which both of them understood the full implications of their consent. It was dictated by Yahweh, and accepted by the Israelites in a state of euphoric bewilderment under the influence of the dramatic and, to them, supernatural happenings at Mount Sinai. They were a simple and trusting people who had no concept of the lengths to which Yahweh would go to ensure their compliance to his will.
>
> The ambiance of the manner in which the Israelite agreement was obtained was a form of duress and, in civilized terms, would have nullified the Covenant when Yahweh's actions, and demands, became oppressive. As far as we know, the common people were never told that one of the requisites of the Agreement was that they should act as Yahweh's troops in the conquest of the Near East.
>
> It makes no difference that the consent of the Elders was obtained during that visit to the mountaintop. Even there, Moses seemed to be surprised at Yahweh's forbearance — 'Yet he did not raise his hand against the leaders of the Israelites; they beheld God, and they ate and they drank.'
>
> The signing of an Agreement would have been followed, traditionally, by food and drink. But why should Moses have even considered that Yahweh might raise his hand against the Elders on such a visit? Surely, only if there had been some altercation, or if there had been reluctance on the part of the Elders, and pressure had been applied to make them sign.
>
> Israel did not ask to be taken out of Egypt, but allowed themselves to be led out in the knowledge that, although unpleasant things were happening to them where they were, worse might follow if they refused to go.
>
> 2. The sanctions proposed by Yahweh, in the event of the Covenant being broken by Israel, were not such as would be acceptable to civilized communities. To threaten fever and consumption; terror; the killing of children by wild beasts; pestilence that would strike the weak and defenceless as well as the strong; reduction to cannibalism through extremes of hunger; and the utter desolation of the

> country, must appall all but the most depraved and power-hungry. In the Curse, there is an essence of vindictiveness and cold-blooded indifference to suffering that is redolent of the worst type of human despot.
>
> 3. The world's major religions all have common factors.... They are based on love, tolerance, justice, care for the weak and suffering, and just rewards for the good life; and, perhaps, a hell of their own making for those who deliberately choose an evil path. But in Yahweh's reprisals, the good were to suffer with the bad; the innocent with the guilty; and the little children, and the frail, with the strong and resistant.
>
> 4. But, perhaps the most disturbing factor of all is that, since those days, history has seen a series of events which have all the trappings of these declared reprisals....[77]

As a result of these observations, the O'Briens conclude with what may be considered "the Gnostic question": "[At] some point... we shall have to consider whether Yahweh was evil: whether the Israelites were right when they cried out in the Wilderness — 'Yahweh hates us!' We shall have to consider whether the Cathars, in the Middle Ages, were right in declaring Yahweh to be the Devil!"[78]

a. Cowing Through Technology and Public Executions

The O'Briens then examine one incident — that of the meeting of the Israelites with Yahweh in the "tabernacle" — from the standpoint of technological displays coupled with public executions, a classic of despotic behavior:

> We visualize the high canvas wall of the east entrance to the courtyard being rolled back to reveal the entrance to the Tent of Meeting. As the hour approached, Elders would have thronged the Courtyard of Assembly; and the mass of the ordinary people would have been in crowded ranks behind....
>
> At last, Aaron raised his hand to still the mounting excitement, and he and Moses went into the Tent of Meeting, in which Yahweh was. The tension in the Assembly must surely have been electrifying: at last they were to see, at close quarters, this Yahweh who seemed

77 Christian and Barbara Joy O'Brien, *The Genius of the Few*, pp. 191–192.

78 Christian and Barbara Joy O'Brien, *The Genius of the Few*, p. 192. Such a view was, the Cathars and Gnostics maintained, behind Christ's statement "You are of your father, the devil."

> part Prince, part Warrior, part Magician, and completely awesome Being — who travelled in a 'cloud,' trumpeted from the mountain-top, and laid down laws of a most exacting kind. And who promised so much prosperity in a barren land!
>
> This Presence was shortly to appear in the doorway. What would they see? Had Moses, and their leaders, been able to convey any sense of the majesty and power of this Being? In the event, the reality must have been more than the expectation: certainly, the staging was superb.
>
> Moses and Aaron came out first, stepped aside and held up their hands for silence. Yahweh appeared in the doorway.
>
> (Lev. 9:23–24): "...and the Presence of the Lord appeared to all the people. Fire came forth from before the Lord and consumed the burnt offering and the fat parts on the altar. And all the people saw, and shouted, and fell on their faces."
>
> This was a highly dramatic introduction to their Lord, who was not above using his technical superiority to hold the attentions of his subjects — or, perhaps, to cow them.[79]

But according to the O'Briens, a mere technological display to cow his subjects was not all on Yahweh's agenda that day:

> Although not described as such, there now appears to have been an interval in the drama. It was a memorable occasion, to be told and re-told in the tents for many generations, and there may have been feasting and drinking with all, including Aaron's sons, joining in the celebrations. And Yahweh must have been looking on, perhaps seated in the doorway to his Tent.
>
> (Lev. 10:1–2): "Now Aaron's sons, Nadab and Abihu, each took his firepan, put fire in it, and laid incense on it; and they offered before the Lord alien fire, which he had not enjoined upon them. And fire came forth from the Lord and consumed them; thus they died at the instance of the Lord."
>
> Up till then, there had been threatenings of death, but these were the first recorded *murders* committed by Yahweh; there were to be many more. But we must be sure that we are justified in pronouncing — *murder*; we must ask ourselves whether there could be any justification for the act.
>
> All the main translations give the same account — so there is no error, that we can see, in the translations. 'Fire came forth from the

79 Christian and Barbara Joy O'Brien, *The Genius of the Few,* pp. 208–209.

> Lord, and consumed them'; and the chronicler assumed that it was a deliberate act. Had the sons of Aaron endangered Yahweh in some way? Or was the killing an act of uncontrolled anger such as Yahweh had warned Moses might happen? — 'if I were to go in your midst for one moment, I would destroy you!' Alternatively, was it a calculated act of execution carried out on two servants who had, unwittingly, stepped out of their place?
>
> A decision is not difficult to reach. Either alternative, by our norms, was completely unjustified; and, in hindsight, after considering the other appalling actions which took place over the time-span of the Wandering, we are forced to pronounce 'Murder'; but we might add — 'while the balance of his Mind was disturbed.' Nadab and Abihu were trying to do honour to Yahweh by presenting him with the aroma of incense; perhaps the flies were thick about him and they sought to bring him some alleviation of the annoyance. But it was still — murder.[80]

This technological prowess was again used, according to the O'Briens, in a far more dramatic way to cow the Israelites, again through what can only be described as "serial public executions."

The episode in question is recounted in Numbers 16:26–35.

> (Num. 16:26–35): ..."Move away from the tents of these wicked men and touch nothing that belongs to them, lest you be wiped out for all their sins." So they withdrew from the abodes of Korah, Dathan, and Abiram. Now Dathan and Abiram had come out and they stood at the entrance of their tents, with their wives, their children, and their little ones. And Moses said, 'By this you shall know that it was the Lord who sent me to do all these things; that they are not of my devising: if these men die as all men do, if their lot be the common fate of all mankind, it was not the Lord who sent me. But if the Lord brings about something unheard of, so that the ground opens its mouth wide and swallows them with all that belongs to them, and they go down alive to Sheol, you shall know that these men have spurned the Lord!'
>
> "Scarcely had he finished speaking all these words when the ground under them burst asunder, and the earth opened its mouth and swallowed them up with their households; all Korah's people and all their possessions. They went down alive into Sheol, with all that belonged to them; the earth closed over them and they vanished from

80 Christian and Barbara Joy O'Brien, *The Genius of the Few*, pp. 209–120.

> the midst of the congregation. All Israel fled at their shrieks, for they said, 'The earth might swallow us!'
>
> And a fire went forth from the Lord and consumed the two hundred and fifty men offering the incense.

The O'Briens' comments are worth citing at length:

> What can we possibly make out of this incredible narrative? The killing of two hundred and fifty men was a form of execution which is becoming commonplace in this account — brought about by flame spurting from Yahweh or, more likely, from some weapon which he held in his hand. This was not a bullet-laden gun, because he had used it to burn up the meat and fat on the range in a peaceable, but dramatic, demonstration of his powers; it was a weapon that generated intense, and localized, heat. Perhaps it was in the same category as the laser equipment which our own technology is now beginning to perfect.
>
> But the swallowing up of the three men's families, with tents and belongings, appears to have been an event of an entirely different kind. Earthquakes in which local tension rifts cause linear openings in the ground several metres across are comparatively common, and there have been cases where subsequent movement has closed a rift, again. But it would ask too much of coincidence for a local tremor to have selected just those people who had incurred Yahweh's displeasure — to have dropped them into the depths and to have closed up, again.
>
> Of course, it would be less of a coincidence if a local earthquake had occurred which swallowed up a portion of the Israeli camp, including some of the insubordinates' families at, or near, the time when Yahweh destroyed the two hundred and fifty men with censers. History, or even a contemporary chronicler, might have connected the two events. Certainly, the area in which the Israelites were wandering was prone, geologically, to tension rifting of the type described. This may have been the explanation, but an alternative should be considered which would be practicable in terms of modern technology, given sufficient energy.
>
> Yahweh appears to have been in possession of a weapon capable of exuding intense heat, possibly to the extent that localized vaporization occurred. It would follow that his aerial craft might have possessed an even more powerful weapon. If such had existed, and had been used selectively to obliterate the families, the path of the beam

> might have left a furrow on the ground which the stunned community might have convinced themselves, later, had been caused by the ground opening and closing.
>
> But whichever cause we attribute to the deaths, we are beginning to be sickened by the carnage apparently needed to correct a 'stiffnecked people,' who had had little or no say in formulating the Covenant that they were being forced to observe. *Indeed, the complete lack of feeling for the suffering of the innocent — wives, elderly people and little children — argues a monstrous megalomania such as has only been attributed to a handful of human leaders in the world's history.*[81]

In other words, the admission or assumption of a technology in play and used by Yahweh to enforce his will on the Hebrews changes the nature of how one reads the text, with the consequences that his "moral" character is no longer so easily rationalizable along the lines of theological piety. More importantly, the admission of such a context also argues that such demonstrations were in part technological, and in part psychological and sociological techniques to keep the Hebrews in a state of fear and compliance. In short, they argue that such incidents — while revelatory of his character — are also possibly demonstrations of mind manipulation techniques by the old tried and true method of all despots: terrorism.

b. An Indefensible Act: Impalings

To drive these points home, the O'Briens refer to one final set of incidents from Numbers 25:3–5 (citing the Jerusalem Bible translation):

> "With Israel thus committed to the Baal of Peor, the anger of Yahweh blazed out against them.
>
> "Yahweh said to Moses, 'Take all the leaders of the people. Impale them for Yahweh here in the sun; then the burning anger of Yahweh will turn away from Israel.' Moses said to the Judges in Israel, 'Every one of you must put to death those of his people who have committed themselves to the Baal of Peor.'

The O'Briens mince no words in what they think of this horrible sentence:

> If we are correct in our interpretation of impaling, this was the most barbarous, and inhuman, form of fatal torture known to man. Its

81 Christian and Barbara Joy O'Brien, *The Genius of the Few,* pp. 219–220, emphasis added.

> description, let alone its practice, should have no place in a civilized document....
>
> We cannot stress too strongly that by this one declaration of intent, Yahweh put himself beyond the pale of civilized behavior; all his cruel acts, all his murders, were insignificant beside this one, appalling judgement. It is true that, as far as we know, there is no record of the order being carried out; but the mere contemplation of such acts is as wicked as their execution. That they should fall within the repertoire of an advanced and self-styled compassionate Being gives reason for a very uneasy concern. And if I were given one wish — it would be that every priest who called on his congregation to worship the 'Great Jehovah' should be called before his Bishop, and instructed to place the real facts concerning this monster in front of his congregation.[82]

The O'Briens are also quick to take issue with Moses' defense for Yahweh's various actions: "Learn from this that Yahweh your God was training you as a man trains his child."[83] However, "it was not an apologia that we would find possible to accept: not sparing the rod is one thing, but wholesale slaughter is entirely different."[84]

5. An Inhuman Face?

What was the source of all this inhumanity? The source, according to the O'Briens, was very simple: Yahweh was not human. Commenting on the fact that Yahweh kept his face hidden from men, they mention two possibilities for the strange behavior:

> (a) that Yahweh kept his face covered, in the close proximity of men, as a protection against human-carried bacilli, or
> (b) that Yahweh's face was so different from a human face that the revealing of it would cause alarm, or distress.[85]

The O'Briens clearly favor the latter alternative, for they note that if one interprets Yahweh as an actual physical being, then his *longevity* over time also accounts for his consistent agenda over several generations and his willingness to extend the Wandering until a new generation of Hebrews had matured, one that had not known life in Egypt, one that would be more compliant.[86]

82 Christian and Barbara Joy O'Brien, *The Genius of the Few,* pp. 225–226.
83 Ibid., p. 230.
84 Ibid.
85 Christian and Barbara Joy O'Brien, *The Genius of the Few,* p. 195.
86 Ibid., pp. 216–217.

6. The Tabernacle: Yahweh's Mobile Palace

But was Yahweh in fact a physical, real being?

To this the O'Briens answer a resounding "yes" and point to the Tabernacle itself as the strongest argument for this view:

> Whatever the reason, the expressed desire for a ground-based dwelling is the most telling factor in the argument in favour of the physical nature of Yahweh — that his craft was not suitable for a lengthy sojourn in the Wilderness.
>
> In our own environment, much can be surmised of a man, or a family, from the house in which they live. Their level of prosperity, their tastes and interests, and even their statures, can be assessed from its appearance and dimensions, and the nature and positions of its furnishings. For example, a highchair would indicate the presence of a young child; and a wall shaving-mirror, by its position, will often indicate the height of the man who uses it.... The writer of the Book of Numbers was well aware of this correlation when he indicated the size of the giant Og, King of Bashan, by giving the dimensions of his bed.[87]

Following this biblical methodology, the O'Briens take note of the sheer *size* of some of the portions of the Tabernacle, particularly those in which Yahweh was said to dwell, or in which he met with the Hebrew leaders. They come to an astonishing, and disturbing, conclusion:

> We have already made it clear that we consider Yahweh to have been a solid physical Being with appetites that required attention like any human; and such a view is consistent with the accounts of the life-styles of the [Anunnaki]. We do not believe that it would be rational to ignore the evidence that points to the Tent being a functional desert dwelling for a Being requiring food and drink, a place in which to work and confer, and a private sanctuary in which to rest. Nor do we wish to avoid the deduction that Yahweh *was a being of exceptional size by human standards.* If the Ark were used for sitting upon; and if the Table were used from cushion level, in desert style, then considering the great height of the Tent walls, Yahweh should have been at least 2.4m (8 feet) tall and, possibly, as much as 4m (13 feet). A height within this range, but nearer the higher figure, would be completely compatible with the Tent, and its furnishings.[88]

---

87 Ibid., p. 196.

88 Christian and Barbara Joy O'Brien, *The Genius of the Few,* p. 202.

Yahweh, on this view, was thus a *giant*, and as giants were, in the Bible, the hybrid offspring of "the sons of God" and human women,[89] the implications of their remarks are clear: Yahweh may himself have been one such offspring!

Taking this method to its logical conclusion, they observe that the appointments of the Tabernacle reveal his opulent tastes,[90] and that its very construction reveals an individual of an exacting nature,[91] while the Tabernacle itself has all the standard characteristics of an actual dwelling.[92] On this view, the Law itself changes, for it is re-secularized as simply the embodiment of Yahweh's idiosyncrasies,[93] and the dietary laws come to be reinterpreted as not cultic, but merely culinary.[94]

7. The Urim and Thummin: More Technology?

As a final argument that Yahweh has access to extraordinary technologies by which he intimidates, and stays in contact with, the Hebrews, the O'Briens mention Aaron's Urim and Thummim, and give a uniquely Sumerian explanation for them:

> By the time of Saul, the terms Urim and Thummim had become so debased that they were used for the casting of lots; they were pressed into use as an oracle in the absence of Yahweh, and our tossing of 'heads' or 'tails' is possibly a relic of this ancient practice.
>
> The important point, here, is that the later Hebrew priests used Urim and Thummim as a symbolic method of communicating with Yahweh, even as dried yarrow stalks, in combination with the I Ching, have been used by the Chinese. At the time of Saul, they may have been small sticks, pebbles or dice (all of these have been suggested) with 'yes' and 'no' connotations, which were picked out, at random, from the pocket of the ephod.
>
> The practice may have been a tribal memory of a more technical method used by Aaron for communication with Yahweh, the instruments of which were no longer available. Following this argument, the use in the desert should have been for direct communication with Yahweh, there being no necessity to cast lots to obtain his decision. Such a communication could have been in two forms: one, to enable

89 Q.v. Genesis 6.
90 Christian and Barbara Joy O'Brien, op. cit., p. 198.
91 Ibid., p. 199.
92 Ibid., p. 200.
93 Ibid., pp. 204–205.
94 Christian and Barbara Joy O'Brien, *The Genius of the Few*, p. 229.

> Yahweh to summon Aaron to the Tent from the father reaches of the Camp and, two, a system by which Aaron could communicate with the aerial craft during Yahweh's periods of absence.
>
> As there is no satisfactory interpretation of Urim and Thummim to be had from Hebrew etymology, there is a case for referring back to the Sumerian syllabic equivalents. A tentative analysis is as follows:
>
> HEBREW: Urim or U'rim
> *SUMERIAN: u* = 'height': *rim* = 'reduce' or 'shorten.'
>
> HEBREW: Thummim or Thum'min
> *SUMERIAN: tum* = 'bring': *min* = 'Shamash' or '20.'
>
> The (Anunnaki) hierarchy were ranked by numbers (tens), and 20 was Shamash's number.
>
> From this analysis, it may be suggested that, in Sumerian, *u-rim* could have meant 'height-reducer' or 'distance-shortener'; similarly, *tum-min* could have meant 'Shamash (or Yahweh)-bringer.'
>
> What weight can be put on this analysis is uncertain; all that can be stated is that it should be borne in mind, as an unsubstantiated speculation, that the Urim and Thummin may have been small, technical devices for effecting communication at a distance, between Aaron and Yahweh. And very useful they would have been. If the Sumerian interpretation should be correct, we could consider the Thummim as a distant alarm call.[95]

Whatever one may make of these Sumerian conjectures, one episode from Numbers 12:1–16 does suggest the use of a technology in play, this time to eavesdrop and, again, to cow the Hebrew leaders. The episode concerns Aaron and Miriam criticizing Moses. At this point, Yahweh, hovering above in his pillar of cloud and fire, immediately orders the three to appear before him in the Tent of Meeting. The technological possibilities are evident from the narrative, according to the O'Briens:

> Miriam and Aaron criticized Moses, and Yahweh heard it and, peremptorily, ordered them to the Tent of Meeting. The narrative has the air of instant action. The culprits were talking outside the Tent, probably away in the camp, somewhere. Yahweh was up in the aerial

95 Christian and Barbara Joy O'Brien, *The Genius of the Few*, pp. 206–207.

> craft, and yet he overheard their conversation and could communicate with them immediately.
>
> Surely, the explanation should like in the fact that it was Aaron who carried the Urim and Thummim. Is it so unlikely that it was through these instruments that communication occurred? Is if not less likely that it was some form of extra-sensory perception on the part of Yahweh? Perhaps, Aaron made an unfortunate mistake; three thousand years later, the President of the United States, in India, was to be the victim of a similar technical error — the microphone that, inadvertently, was left switched on. Or, perhaps, in Aaron's case, the microphone could not be switched off; perhaps he was permanently 'bugged' without being aware of it.[96]

Given all the assumptions of a technology in play and their analysis and understanding of Yahweh's character, bugging Aaron would be consistent with such an individual and his desire to maintain a tight grip on the intermediaries between him and the Hebrew people, and to stamp out dissention.

8. The Shining Ones and the Possible Agendas

So what, at this point, do we have? And how does all this analysis of Yahweh and his behavior fit into the context of mind manipulation? And what, if any, agendas does it imply? In order to understand the O'Briens' argument it is first necessary to understand that they *assume* the basic historicity of the account that they reinterpret. Once this is done, their argument boils down to four points:

1) Yahweh is a physical being, not a spiritual one, or a deity in the conventional sense;
2) As such, Yahweh is also apparently a long-lived being, whom they interpret in consonance with similar long-lived beings in the ancient Sumerian tablets. This longevity accounts for his willingness to forego immediate conquest of Canaan in favor of letting the older generation "die off";
3) Assuming his physicality, they argue that the Tabernacle is most fit for a being of very large stature, i.e., a giant, one of the very offspring of the "sons of god and daughters of men" spoken of in

96 Christian and Barbara Joy O'Brien, *The Genius of the Few*, p. 215. The Sumerian connection may not be as tenuous as the O'Briens suggested, for one need only recall Delitzsch's dilemma and the presence of the name Yahweh itself — in the form of Iave — in Sumerian cuneiform tablets.

Genesis 6, and that as such, Yahweh is possibly one of the "fallen Watchers" that will be dealt with in this book in part two;

4) Yahweh uses his technologies and his behavior to cow the Hebrews into unquestioning obedience of his orders, no matter how they might go against the normal urges of human conscience.

With this in mind, it is worth taking a closer look at Yahweh's behavior in the light of some of the *techniques* of mind manipulation, for therein is disclosed the possibility that it was nothing but an agenda all along.

9. "Reverse Depatterning," Psychic Driving, and the Stockholm Syndrome

The behavior of Yahweh, viewed in the light of the techniques of mind manipulation, resemble nothing so much as those of (1) a "reverse Depatterning" based on sleep and rest deprivation, (2) "Psychic Driving," and (3) the well-known Stockholm Syndrome.

"Psychic Driving" and "Depatterning" are the rather euphemistic names given to two techniques of mind manipulation by their "inventor," psychiatrist Ewen Cameron, during his secret work for the CIA's top secret MKULTRA mind control program.[97] Cameron, known to his colleagues as an impatient man, had little use for the time-consuming methods of Freudian psychotherapy to achieve results in patients.[98] Instead, he developed a technique that he called "depatterning," designed to cure schizophrenic patients of the patterns of schizophrenic behavior, if not cure them of the disease itself. This treatment consisted of endless days — usually up to 30 — of endless sleep, induced by a potent cocktail of sedatives administered several times a day, along with a regimen of electroshock "therapy" that was administered at least three times a day.[99] His goal was to induce essentially a *tabula rasa* in the mind of his patients, which could then be remolded with behavioral patterns derived from "psychic driving."

For as much as 16 hours a day, while patients lay in a drug-induced "stupor" again brought about by a potent cocktail of sedatives, patients would hear messages played over and over, repeatedly, as they tried to rest. This began with several weeks' "treatment" of this process, with the initial stage being a

97 It should be stressed that there is some question of whether or not Cameron knew he was actually doing work for the CIA, though the ultimate source of some of his funding was indeed from the agency. See John Marks, *The Search for the "Manchurian Candidate": The CIA and Mind Control: The Secret History of the Behavioral Sciences,* pp. 141–142. It is also to be noted that Cameron's "research" was also funded in part by the Rockefeller Foundation, q.v. p. 141.

98 Ibid., p. 141.

99 John Marks, *The Search for the "Manchurian Candidate",* pp. 140, 142.

"negative" stage where the patient was played messages reinforcing their failures and guilt.[100] After this "guilt reinforcement" phase, Cameron would then switch them over to another two to five weeks of "positive reinforcement," utilizing the same drug-induced stupor and repeated recorded messages designed to induce new patterns of behavior.[101]

It takes only a little imagination to see how in the hands of some religious cults a kind of "reverse depatterning" and "psychic driving" occurs, utilizing not drug-induced sleep, but rather the reverse: intense sleep deprivation and prolonged exertions of physical work coupled with always being under the scrutiny and never out of the sight of the "depatterner," with the slogans of the "cult leader" repeated over and over again to the dazed victims, reinforcing their unquestioning obedience in constant repetition, like a religious litany almost, and his "love" for the cult followers in spite of the complete lack of actual normal demonstrations thereof. In either case, the methods are rather similar to the methods recorded by the first books of the Old Testament: a relentless driving of a whole people, a breakdown of will, endless rehearsals of love-and-threats, the end result of which is similar — a breakdown of normal patterns of behavior and emotion.

There is a final suggestive sign that perhaps these or similar techniques were in play by Yahweh toward the Hebrews, and that is the well-known Stockholm syndrome. According to the article on Stockholm syndrome in the online encyclopedia Wikipedia, there are four conditions that are the *sine qua non* for the syndrome to occur in its victims:

- Hostages who develop Stockholm syndrome often view the perpetrator as giving life by simply not taking it. In this sense, the captor becomes the person in control of the captive's basic needs for survival and the victim's life itself.
- *The hostage endures isolation from other people and has only the captor's perspective available.* Perpetrators routinely keep information about the outside world's response to their actions from captives to keep them totally dependent.
- *The hostage taker threatens to kill the victim and gives the perception of having the capability to do so.* The captive judges it safer to align with the perpetrator, *endure the hardship of captivity, and comply* with the captor than to resist and face murder.
- *The captive sees the perpetrator as showing some degree of kindness.* Kindness serves as the cornerstone of the Stockholm syndrome;

100 Ibid., p.145.
101 Ibid.

> the condition will not develop unless the captor exhibits it in some form toward the hostage.... if perpetrators show some kindness, victims will submerge the anger they feel in response to the terror and concentrate on the captors' "good side" to protect themselves.[102]

It takes little imagination to see the key elements of the syndrome in the techniques and technologies — the public executions with reassurances of his love — displayed by Yahweh.

Does all of this point conclusively to an agenda in play?

Of course not. However, if one grants the proposition of a surviving elite from that ancient war, then religion would surely be one of the techniques employed to create suitable climates of opinion, and certainly one of the techniques and tools used to embark upon campaigns of wholesale slaughter and conquest in the name of a "god" who sanctioned it. At the minimum, viewed in this admittedly secular light, the techniques displayed by Yahweh in the books of Exodus and Numbers have about them the all too familiar and haunting similarity to the techniques used by modern mind manipulators and cult leaders. They are abusive behaviors, as a friend put it to me recently, that would land such a tyrant in jail today.[103]

Or worse, perhaps before a war crimes tribunal for crimes against humanity. Whatever one makes of the views of the O'Briens, one thing cannot be ignored, and that is, they have raised the bar of the task of religious apologetics considerably.

---

102 Wikipedia, *Stockholm Syndrome,* citing the *FBI Law Enforcement Bulletin,* July 2007, www.fbi.gov.publications/leb/2007/july2007/july2007leb.
htm#page10, emphasis added.

103 There might be an editorial agenda in play, as suggested by the alternative researcher William Bramley: "...the tale of Noah, like many other stories in the Old Testament, is taken from older Mesopotamian writings. Biblical authors simply altered names and changed the many 'gods' of the original writings into the one 'God' or 'Lord' of the Hebrew religion. *The latter change was an unfortunate one because it caused a Supreme Being to be blamed for the brutal acts that earlier writers had attributed to the very* ***un****-God-like Custodians."* (William Bramley, *The Gods of Eden* [Avon Books, 1990], p. 46). Of course, that may have been the *point* of the change, to claim to be the Supreme Deity precisely in order to exact unquestioning obedience, and thereby to justify immoral acts with the cloak of "morality" obtained by "obedience" to someone who, in the end, was an imposter.

Four

Elite with Agendas:

Conclusions to Part One

∴

"At length this universal knowledge was diuvided into several parts, and lessened in its vigour and power. By means of this separation, one man became an astronomer, another a magician, another a cabalist, and a fourth an alchemist. Abraham, that Vulcanic Tubalcain, a consummate astrologer and arithmetician, carried the Art our of the land of Canaan into Egypt..."
—Paracelsus[1]

CHIMERICAL CREATURES THAT disconcertingly appear on the Ishtar Gate of Babylon in the context of other, very real, creatures; Hebrew names of God that equally disconcertingly appear in cuneiform tablets long before the name was supposedly known; elites with an agenda to impose an astronomically- and geodetically-based system of measure over a widespread area; mind manipulators, torsion temples, and a god of "love" that appears to know some very elaborate techniques of mind manipulation: what tentative conclusions may be drawn from it all?

In *Babylon's Banksters* I posed the following problem:

> (Alexander Del Mar) states that "The governments of Persia, Assyria, Egypt, Greece and Rome made a profit on the coinage by raising the

1 Paracelsus, *The Aurora of the Philosophers,* from *Paracelsus, his Aurora & Treasure of the Philosophers, As also the Water-Stone of the Wise Men; Describing the matter of, and manner how to attain the universal Tincture, Faithfully Englished, and Published by J.H. Oxon.* London, Giles Calvert, 1659.

> value of gold, while those of India, China, and perhaps also Japan, made their profit by maintaining, or enhancing, the value of silver."[2] In other words, for the societies of the Occident — Egypt, Assyria (and presumably Babylon), Persia, Greece, and Rome — artificially defined gold as being the metal of highest value in terms of its convertibility into more units of other metals, while, conversely, the governments of the Orient — India and China — pursued the reverse policy, of making silver the highest valued metal in terms of its convertibility into other metals. Thus, trade could be carried out between these two disparate parts of the world, the policies were in a certain sense an inevitable consequence of that trade. However, a closer examination reveals a hidden player, for such trade will inevitably create the rise of an international trading class, one which, moreover, will create its wealth precisely by trade in these precious metals, metals that are easier to transport than finished goods, and which can be exchanged in any place for such goods. In short, what is being created, from earliest times, is an international financial class of "bullion brokers," or as we would call them now, bankers. A significant question now occurs: *Is it possible that, rather than such a class having emerged as a consequence of such governmental policies and trade, that the converse is true? Is it possible that there existed such a class of "international bullion brokers" who* ***created*** *these policies in various parts of the world, policies which would enhance their own power and wealth?* If so, then how did they achieve and orchestrate this?[3]

With the research of Knight and Butler in hand — with their own conclusions that such an elite existed — then we are now in a position to answer those questions with a decisive "yes." But there is a caveat: to assume the existence of such an elite in megalithic times, while certainly indicated by the research of Knight and Butler, does not prove the existence of the same elite engineering bullion policies millennia later, much less does it prove the continuity between the two.

There is, however, an important factor to bear in mind here, one indicated by the ancient texts themselves, from the Sumerian Kings' list, to the genealogies of the Torah, to the Egyptian texts, and that is the extraordinary longevity claimed for those "gods" who came down and midwifed human civilization into existence. Under such conditions, and even allowing for the possibility that the alleged longevity of these "gods" is greatly exaggerated, one is none-

---

2 Alexander Del Mar, *History of Monetary Systems,* p. 89.

3 Farrell, *Babylon's Banksters,* p. 162, italicized emphasis added, bold emphasis in the original.

theless left with the possibility that under conditions of longevity of even hundreds of years, it would be much *easier* for such an elite to maintain cohesion of personnel and ideological outlook through such a length of time. The assumption is less difficult than one might imagine, for as Knight and Butler observe, the establishment of accurate weights and measures was a necessary step to the establishment of international trade, and thus, the *activity* between the two poles separated by some millennia is *consistent*, and this would suggest that one is indeed in the presence not only of a continuous ideology and policy, but in the presence of a continuous elite.

If these speculative propositions be granted, then the question I posed in *Babylon's Banksters* is answered, though the answer is an extraordinarily disturbing one: *the elite preexisted by millennia the bullion policies that came to be in evidence with the rise of private monopoly issuance of money as debt and the bullion policies of the states issuing it in classical times.* In that context, one must allow for another possibility that was also outlined in *Babylon's Banksters,* namely, that the subsequent rise of an elite *opposed* to the state issuance of money as debt-free receipts on the surpluses of the state warehouse might not represent the rise of a *new* elite with a new agenda, but rather, the final break of an elite that existed coterminously with the one behind the establishment of weights and measures, and with which it made common cause for a period. But howsoever one slices it, one does appear to be in the presence of at least *one* such elite with a consistent agenda over a great period of time that perhaps fragmented later into two opposing ones, or perhaps two elites at the outset, making common cause for a time.

There is another factor in play, and that is, at the end of this period when one discovers this elite — or these elites — within the religious temples of antiquity, and involved in the issuance of money. Very simply put this means that this elite — or elites — is also involved in the manipulation of *religion*, and thus, it is involved simultaneously with the manipulation of *two* powerful institutions for the creation and maintenance of social cohesion: finance and religion.

In the case of the temple, we have seen that at least *some* rudimentary technology of communication existed in the form of *some* of those ancient temples, raising the possibility that many of them might have been covertly conceived for that purpose. Again, this is a likely activity for such an elite to undertake, both to manipulate and coordinate financial policy over large distances, and to manipulate and control "revelations" and "oracles." In the case of Yahweh, we have seen disturbing indications of at least a *technique* of mind and social manipulation in play reminiscent of the Stockholm syndrome and of other techniques explored by the CIA, if not of an outright technology (a debatable proposition on any view).

Which brings us back to Koldewey's conundrum, for one of the things he noted in his explorations of the "Sirrush" mystery was that apparently it had to be a real creature because the priests were tending it. Given the apparent chimerical nature of this creature — if real — this suggests that there may have been another possible agenda going on as well in ancient times: genetic manipulation...

II.

Genes and Giants, or,

If "It" is in the Genes, then what is "It"?

"Human beings appear to be a slave race languishing on an isolated planet in a small galaxy. As such, the human race was once a source of labor for an extraterrestrial civilization and still remains a possession today. To keep control over its possession and to maintain Earth as something of a prison, that other civilization has bred never-ending conflict between human beings, has promoted human spiritual decay, and has erected on Earth conditions of unremitting physical hardship. This situation has existed for thousands of years and it continues today.

"Furthermore, it is conceivable that the alleged ownership of Earth may have changed hands over the millennia, in the same way that ownership of a corporation can pass among different owners without the public being aware of it."

—William Bramley, *The Gods of Eden,* pp. 34, 37.

Five

The Genome Wars:

Modern and "Mesopotamian"

∴

"It wasn't hard to see that once you can slice and dice DNA and clone quantities of just the bits you want, the possibilities are endless: you can manufacture proteins in bulk, engineer crops with built-in insecticide genes, introduce healthy genes into patients who lack one needed to survive — in essence, you can redesign life."

—James Shreeve[1]

"In genetic terms, this mixture was to be half Lordling and half Human; and since the former are stated to have provided the male elements, the female elements must have been taken from human women..."

—Christian and Barbara Joy O'Brien[2]

THERE WERE, both in modern times, and in ancient ones, what may best be described as "genome wars," fought over who would control the very code of life itself: the public sector, or the private one. While the modern debate raged over ethical issues and the rights of private corporations to patent various genes and processes, the race itself pitted two powerful minds —

1 James Shreeve, *The Genome War: How Craig Venter Tried to Capture the Code of Life and Save the World* (New York: Ballantine Books, 2004), pp. 99–100. This book is probably the best and most readable account of the race between the public Human Genome Project and Venter's private company, Celera, to map the entire human genome. For a more technical, though dated, approach, see Robert Cook-Deegan, *The Gene Wars: Science, Politics, and the Human Genome* (New York: W.W. Norton and Co., 1994).

2 Christian and Barbara Joy O'Brien, *The Genius of the Few*, p. 161.

the public Human Genome Project's Francis Collins, and the private corporation Celera's Craig Venter — in a race that came, quite literally, right down to a brokered "tie" engineered by the administration of former American President Bill Clinton.

Here, as elsewhere, the modern lessons afford a template by which to interpret the records of the past, for a similar race was apparently held in ancient times, and if the lessons of the contemporary one are any indicator, the ancient results are rife with a potentially horrifying implication. But before we can state clearly what that implication is, a closer, though necessarily cursory look, at the modern race and all its implied technologies and legal issues is in order.

A. An Overview of the Modern Genome War: The Race between the Human Genome Project and Craig Venter's Celera Corporation

"In hindsight, it is hard to imagine there would have been a race to sequence the human genome without Craig Venter."[3] While it is impossible and indeed unnecessary to recount in detail the race between Craig Venter's private Celera Corporation and the public Human Genome Project headed by Dr. Francis Collins, some brief overview of that race is necessary, if nothing else than to highlight the magnitude of the task, and the technologies and techniques that eventually accomplished it, for these in turn will provide a basis to unlock and possibly decode some astonishing assertions in some very ancient cuneiform tablets.

The genome is often described as "the book of life," and indeed, the analogy of a book is quite apropos to describe the way the DNA helix works. "Imagine," says author Matt Ridley, "that the genome is a book."

> There are twenty-three chapters, called CHROMOSOMES.
> Each chapter contains several thousand stories, called GENES.
> Each story is made up of paragraphs, called EXONS, which are interrupted by advertisements, called INTRONS.
> Each paragraph is made up of words, called CODONS.
> Each word is written in letters, called BASES.
>
> There are one billion words in the book, which makes it longer than 1,000 volumes the size of this one, or as long as 800 Bibles. If I read the genome out to you at the rate of one word per second for eight hours a day, it would take me a century. If I wrote out the

3 James Shreeve, *The Genome War*, p. 117.

> human genome, one letter per millimeter, my text would be as long as the River Danube. This is a gigantic document, an immense book, a recipe of extravagant length, and it all fits inside the mircroscopic nucleus of a tiny cell that fits easily upon the head of a pin.[4]

To appreciate the magnitude of mapping the entire human genome, we need to understand the "bases" that comprise the "words" or codons. These words are composed of never more, and never less, than three bases or "letters," A, C, G and T, which stand for the proteins adenine, cutosine, guanine, and thymine. The basic "grammar" of these letters is that A pairs only with T, and G only with C.[5]

Thus, "to make a complementary strand therefore brings back the original text. So the sequence ACGT becomes TGCA in the copy, which transcribes back to ACGT in the copy of the copy. This enables DNA to replicate indefinitely, yet still contain the same information."[6] But in the midst of all this microscopic complexity, there lurks a mystery, and it may be a significant one for our purposes. As James Shreeve puts it, "There are a lot of extra letters in the genome, sloppily referred to as 'junk DNA,' which do not spell out protein recipes but may serve some other purpose, perhaps vital, perhaps not."[7] In other words, a significant portion of the human genome contains genes for which biologists and geneticists cannot divine any function whatsoever. Indeed, what was so unusual about the human genome as distinguished from any other species was the sheer *amount* of this "junk DNA," for the human genome, consisting of over a billion such "letters," was mostly comprised of this "so-called junk, possibly without any biological purpose at all."[8]

But as James Shreeve observes,

> "Junk" is a misnomer: although protein-coding genes account *for less than 3 percent of the DNA in the human genome,* inferring that the rest is worthless is like saying there is no value in the deserts of the Middle East because they are composed mostly of sand and only a little bit of oil. The fact is, we don't know what purposes lie hidden in that alleged junk. We do know, however, that some of it performs the vital function of regulating when a gene is turned on or off. Without those

4 Matt Ridley, *Genome: The Autobiography of a Species in 23 Chapters* (New York: Harper Perennial, 2006), p. 7.

5 Ibid., p. 8.

6 Ibid., pp. 8–9.

7 Shreeve, *The Genome War,* p. 15.

8 Ibid., p. 40f.

> switches, there would be no difference between a liver cell, a brain cell, or a cell in your big toe, and we would all be a dysfunctional chaos of overexpressed [sic] protein.[9]

In other words, the so-called "junk DNA" functioned as a kind of "computer algorithm" telling the rest of the code when to execute certain functions in the program, and when not to. But that still left the all-important questions, *where did it come from?* Why is there so *much* of it in the human genome by comparison to other species?

But "junk DNA" played an important role in the "genome war," for it was precisely because of these "regulatory regions" that James Watson, co-discoverer of the double helix with Francis Crick, decided to go after the entire sequence of the human genome.[10] The enormity of the task, however, meant that the project would take — or so the thinking ran at that time — a great deal of time and effort, years, if not even almost a decade.

Enter Dr. Craig Venter, and his private Celera Corporation, founded for the express purpose of mapping the entire human genome. In May of 1998, Venter announced that with the financial backing of the Perkins-Elmer Corporation, he was founding Celera (from the Latin word for "speed"), a "private company to unravel the human genetic code." Venter announced that he planned to complete the entire project in the unheard-of time of a mere three years![11] It was a bold, perhaps even brazen announcement, for "nothing like the particular scheme he was proposing had been attempted before. If it were broken down into its various technical components, most of them had never even been attempted before, either."[12] In essence, if one wishes to compare the initial strategies of the public Human Genome Project and Venter's private Celera venture, the aim of the former was quality, whereas Venter's aim was speed. Thus, the Human Genome Project's early strategy was to map each individual gene first, and then assemble the pieces — like a gigantic jigsaw puzzle — into their proper sequence later.[13] Venter's goal was much more ambitious, for not only did he wish to map every single human trait,[14] but, by using massive amounts of DNA-sequencing machines in a Manhattan Project-sized assembly line that would blast the DNA into millions of tiny segments, reassemble and sequence the entire "book" of human DNA using supercomputers and very complex computer algorithms to reassemble the

9 James Shreeve, *The Genome War,* p. 80, emphasis added.

10 Ibid., p. 80.

11 Ibid., p. 6.

12 James Shreeve, *The Genome War,* p. 6.

13 Ibid., p. 70.

14 Ibid., p. 170.

pieces of the jigsaw in their proper order. It was this "shotgun" approach of Venter that called forth rounds of denunciation from scientists within the public project,[15] and yet that galvanized its leader, Dr. Francis Collins, to recentralize what up until then had been a variety of public laboratories and university efforts into a more coordinated effort,[16] and that also caused him to re-evaluate the basic strategy the public project was pursuing. After Venter's announcement, the public Human Genome Project adopted a mediate strategy between its initial "qualitative" approach and Venter's shotgun approach, determining that it would go after a "rough draft" of the human genome sequence.[17]

However, for the public project, there was a fly in the ointment, and that fly was the "Bermuda Accords," to which all participants in the public project had subscribed. By mutual consent, all participants in the public Human Genome Project had agreed that, once individual pieces of data — the bits of the "jigsaw map" — had been completed by the project, these data would be made publicly accessible to everyone. This meant, of course, that Venter's "Celera could grab their data off the web like everyone else." The faster the public project went, "the faster their enemy could go."[18] This placed the Human Genome Project in a Catch-22.[19]

By adopting a kind of "Manhattan Project" approach using massive numbers of DNA sequencers and supercomputers with complex algorithms to assemble the pieces, Venter had in fact, reversed the initial roles that the public and private projects had assumed. After a few months into the race, Venter's Celera was in fact pursuing a detailed quality map of the entire genome, while the public project was aiming for a "rough draft."[20]

In the end, the race was so close that the Clinton Administration stepped in, and brokered what can only be described as a "truce" between Collins' public Human Genome Project and Venter's private Celera corporation in a declared "tie."[21]

---

15 Ibid., pp. 21, 26, 184.

16 Ibid., p. 126.

17 Ibid., p. 188.

18 James Shreeve, *The Genome Wars,* p. 191. Q.v. also p. 198.

19 Ibid., p. 198. Celera was caught in its own Catch-22 as well, for its database was made available to drug companies and private bio-research companies for a subscription fee, yet the Human Genome Project, by making its data publicly available for free undercut Celera's potential customer base. (q.v. pp. 212–213). Thus, Celera, in order to win customers, had to move that much faster than the public project, only adding more fuel to the race between the two.

20 Ibid., pp. 204, 338, 341.

21 Ibid., pp. 343–344. Obviously, this brief survey has left out massive twists and turns and political scheming on both sides. Shreeve's book is by far the *best* survey of the entire "genome war" and is, in a word, magisterial and essential reading for anyone wishing to understand the details of that war.

B. Technologies and Legalities

For our purposes, it is the technologies, techniques, and legal ramifications of the genome project and genetic engineering that must hold the center of our attention, for these three things provide the interpretive key by which to pry loose possible hidden meanings in some very ancient texts. The task of sequencing the enormous amount of information coiled up in the human double helix was the most daunting scientific problem mankind had ever faced, for even if every gene was "decoded," the problem of fitting a billion pieces of data into a coherent map would require not only massive amounts of DNA sequencing machines, but massive amounts of computer power, plus a computer program able to assemble the data spit out by the sequencers into a coherent picture. The benefits, however, were well worth the effort, for with the ability to sequence human DNA, any other organism was, by comparison, a comparative "snap." Moreover, with such maps in hand, one might be able to derive *genetically-based definitions of life itself*,[22] or even to figure out the minimum amount of genes required in order for there to *be* life,[23] and finally, detailed genetic knowledge of organisms would conceivably resolve one of the thornier problems within biology itself: taxonomy, or how to classify various species or even to determine if something *was* a distinct species.[24] Additionally, as we shall see, the techniques of genetic engineering required accurate maps of the genome. Once that map was had, the implications — both good and ill — multiply like rabbits. So, a closer look at the technologies, techniques, and legal implications is in order.

1. The Technology: Sequencers

To appreciate the revolution that Venter's private Celera corporation brought to the *speed* of unraveling the genome, one must understand the state of sequencing technology prior to it. The initial technique for reading a DNA sequence was developed by Fred Sanger of Cambridge University, and to appreciate its complexity and difficulty, it is best to cite Shreeve's description of it at length. It was an entirely manual technique, and therein lay the problem:

> The difficulty of the problem was much greater than the metaphor of "reading letters" suggests. Unlike letters, the four nucleotides of DNA cannot be distinguished by their shape. Sanger's method required a

22 James Shreeve, *The Genome War*, p. 110.

23 Ibid., p. 112.

24 Ibid.

> great deal of ingenuity. He began by preparing a solution containing millions of copies of a small DNA fragment, and divided the solution up into four equal parts. Then he heated the four test tubes, which separated the double-stranded DNA into single strands. To each tube, he added DNA polymerase, an enzyme that uses a single strand of DNA as a template to re-create its missing partner, a complementary string of base pairs. He also introduced a primer — a small synthesized fragment of DNA of a given sequence of nucleotides. The primer fragments would glom onto their complements on the template, which would tell the enzyme where to start the copying process. He also supplied each tube with plenty of free-floating nucleotides: the As, Ts, Gs, and Cs that the enzyme would need as raw material to construct the complementary strings.
>
> At this point in the experiment, the solution in each of the test tubes was exactly the same. Next came the ingenious part. Sanger spiked each tube with a smaller amount of a doctored form of one of the four nucleotides, which had been tinkered with to stop the copying process in its tracks. In the first tube, for instance, there might be a generous supply of ordinary Ts, As, Gs, and Cs but a few of the doctored dead-end Ts as well. Whenever the enzyme happened to grab one of these killjoys and attach it to the growing strand, the reaction on that particular strand would cease. Thus, after reheating the solution to separate the double-stranded DNA again, Sanger ended up with a collection of single strands of differing sizes in that particular tube, each one beginning at the start of the sequence, and each ending when a killer "T" was attached. The same process was going on simultaneously in the other three tubes with the three other DNA letters.
>
> To read the sequence of the entire original DNA fragment, Sanger thus had only to sort the fragments by their size, and read the last letter of each one.[25]

But the method was obviously cumbersome, for with such a technique, mapping the human genome "would take 100,000 years, which alone might explain why a lot of very smart people initially thought the Human Genome Project a very stupid idea."[26] Clearly some sort of engineering revolution was needed if the approach was to be less cumbersome and capable of more rapid sequencing.

---

25 James Shreeve, *The Genome War*, pp. 157–158.

26 James Shreeve, *The Genome War*, p. 159.

Enter Tim Hunkapiller and his revolutionary idea.

In the technique developed by Sanger, "the DNA letters were read by the researcher's eyeball, scanning both across and up the gel at the same time, a process that was both error-prone and squintingly slow."[27] What was needed was a process that would mechanize the entire method.

> Tim's idea was to color-code the last letters in each fragment by chemically attaching a different-colored dye to each of the four doctored, killjoy nucleotides: blue for C, yellow for G, red for T, and green for A. As each fragment in turn arrived at the bottom of the gel, the color of its last letter could be read by a detector, which would feed this information to a computer. If the process worked, several samples could be run at once.... Theoretically, it could sequence more DNA in a day than a single researcher could do in a year.[28]

Yet this meant that the gels themselves still had to be manually prepared on slides. While the major revolution had indeed been achieved, something else was still needed if Venter's goal of mapping the entire genome within three years was to be feasible. The problem with preparing such slab gels was simply that one always ran the chance of one bleeding into another, thus ruining the whole experiment.

> An alternative to slab gels had been in the air for years. Instead of running dozens of samples down a common gel, each sample might be enclosed in a thin, gel-filled capillary tube, where it couldn't possibly wander into its neighbor's lane. The capillary technology depended on the manipulation of incredibly tiny samples of DNA, however, and whenever things get very tiny, the margin of error shrinks in proportion.... Hunkapiller also had a team, sworn to secrecy, working on a multi-capillary machine, code-named the Manhattan Project.[29]

Eventually this project was successful, and it was these machines that Venter employed *en masse* — and the public project as well — to bring about a much swifter conclusion to the project than anyone initially imagined.

---

27 Ibid., p. 60.

28 Ibid.

29 James Shreeve, *The Genome Wars,* p. 64.

2. *The Technology: Super-Computers*

Sequencing the entire human genome into an accurate map required not only the computing power of a supercomputer, but a whole new type of computer *architecture.*

> A typical supercomputer — say, a Cray — built its muscle by stringing together enormous numbers of processors. But the all-at-once assembly of the human genome could not be solved by lots of little processors; it required gross amounts of active memory that could handle a single process, very fast. Virtually all computers on the market at the time employed a 32-bit architecture, which was limited by physics to 4 gigabytes of active memory. The assembly algorithms alone were estimated to need 20 gigabytes of RAM, at the same time that the computer would be servicing all the rest of (Celera's) needs. Compaq had won the Celera contract because its new supercomputer, the Alpha 8400, was built upon a 64-bit architecture that could handle 128 gigabytes of RAM — four thousand times the active memory of an average desktop machine. (Celera) had tested the Alpha 8400 and a competing machine from IBM by seeing how long it took them to assemble the DNA fragments from the H flu genome. Four years earlier, TIGR's 32-bit Sun mainframe had completed the assembly in seventeen days. IBM's new supercomputer finished the compute in three days, fifteen hours. Compac's Alpha 8400 took only eleven hours. Celera ordered a dozen Alphas, to start.[30]

All this added up to the fact that, at that time, ca. 1998–2000, Celera had the largest supercomputers in private hands, outdone only by the facilities at the U.S. Department of Defense as Los Alamos.[31]

3. *The Technique: Computer Algorithms*

In addition to the massive computational power required for the accurate mapping of the genome, there was also a massive *software* problem as well. The sheer *size* of the computer program itself required to do so would be immense, not to mention the several discrete *tasks* that it would have to perform, and to perform flawlessly, for the fact was that "a computing problem the size

30 James Shreeve, *The Genome War,* p. 167.
31 Ibid., pp. 172–173.

of the entire human genome had never been attempted before."[32] Indeed, it could honestly be said that the problem of accurately sequencing the entire human genome was as much a triumph of computer programming as it was of biology or revolutions in sequencing technology.

The initial programming difficulty lay in these two precise things: (1) the sheer size of the genome itself and the amount of data that needed to be processed and assembled, and (2) the discrete tasks that such a complex program would have to perform. It could not, therefore, be reduced to a single principle or mathematical formula.[33]

> The sequenced fragments of DNA coming out of the machines are the puzzle pieces, and the genome is the image they form when all the pieces are locked into place. But there are crucial differences. First, jigsaw puzzle pieces lock together by the shape of their edges, while each piece of a genome puzzle must *overlap* its neighbor by a handful of base pairs — around fifty, to be statistically sure — in order for them to be candidates for a match. Second, a jigsaw puzzle's pieces are all carved up (but) form a single image, so each one is unique. The fragments of DNA to be assembled aren't unique at all. On the contrary, each one *must* contain bits of sequence that are also represented on other pieces, or else there would be no way to overlap them, and no way to put the picture together.
>
> The upshot of this is that one needs a whole lot of pieces in order to put the picture together.[34]

Even assembling the puzzle of the lowly fruit fly, with its only 120 million base pairs, required some six trillion calculations by a computer![35]

Celera's programmers "visualized tiny islands of sequenced code like growing crystals — far-flung stars in a great dark sky gradually extending their perimeters toward their distant neighbors until, finally, they touched."[36] But this only highlighted the multitude of discrete tasks that the program would have to do, not to mention its sheer size:

> ...(The) algorithm would actually entail a series of discrete problems, each of them fanning out in an array of subprocesses, and those subprocesses would be further broken down into their own components,

32 Ibid., p. 138.
33 Ibid., p. 142.
34 James Shreeve, *The Genome War*, pp. 144–145.
35 Ibid., p. 145.
36 Ibid., p. 147.

> and so on, until one reached the level of the hundreds of thousands of individual lines of computer code that would have to be written to make the whole thing work. Ways of checking and refining each component would be incorporated into the program. But whether it worked or not would be revealed only in the last step, when the time came to upend the can and see what tumbled out.[37]

In addition to all *this,* there was also the problem of "statistical noise," i.e., the accuracy of the data coming from the biologists' sequencing machines. Thus, the program also had to be designed in such a fashion that it could work with realistic data coming from the sequencers.[38]

Yet another problem facing Celera's programmers was the problem of contiguous DNA fragments and the whole phenomenon of having to fit the "DNA islands" together by looking at overlapping base pairs. The "Overlapper" component of the computer program, however, faced a fundamental difficulty, and that was that

> ...both the (fruit fly) and human genomes were riddled with identical repeating sections. As a consequence, all too often a fragment of DNA would overlap with two, three, or even a dozen other pieces, because the overlapping sequence was represented in more than one place in the genome. At all costs, the assembly program had to avoid making false connections, which could mislead researchers for decades to come. For the next step of the assembler, then (Celera's programmer) Myers had written a program that essentially broke apart most of what the Overlapper had put together, keeping only those joins where a piece went together with only one other piece. Myers called the result of these uniquely joined pieces a "unitig." In effect, rather than attacking the problem of a complex genome's thousands of repeats, the Unitigger stage dodged the issue for the moment, telling the computer, "Let's just assemble the pieces we *know* go together correctly and throw the rest into a bin to deal with later." The Unitigger was an act of genius, in one stroke reducing the complexity of the puzzle by a factor of 100.[39]

In short, the "Celera Assembler," which is what the program was eventually called, was confronted, via the overlaps, with "an infinite chain of forks in a

---

37 Ibid.

38 James Shreeve, *The Genome War,* p. 173.

39 Ibid., pp. 175–176, emphasis in the original.

path, and the program had to be able to choose the right turn every time and not overlap one piece of code with an extremely similar one that might actually be miles away in the genome."[40] After testing the program against several already known or well-mapped genomes, Celera's program was finally successful, and its lawyers moved immediately to patent the program.[41]

4. The Legal Implications

The patenting of Celera's Assembler is the gateway into the larger issue of patenting specific genes and engineered life forms. In a way, the rush to patent various genes or genetically engineered life forms might be considered a kind of genetic or "molecular land grab."[42] Early on, in fact, an isolated human gene was ruled by the U.S. Patent and Trademark office that it could be considered "valid intellectual property, *if it fulfilled the requirements that any other invention had to meet in order to get a patent.*"[43] It is important to note what is being said here. The entire genome of an organism, since it already exists in nature and cannot be invented, cannot be patented. It would be like trying to patent an eagle or an oak tree or, for that matter, a mountain range. But one could patent certain individual genes or sections of the genome if they were separated for some function "by the hand of man."[44]

Under this understanding then, there are four requirements in U.S. patent law that an invention must meet to be patentable intellectual property:

> The invention must first be original. It cannot have been published before, or be too much like some previous invention. Second, it must be "nonobvious." You cannot get a patent by wrapping a rock in cloth and calling it a no-scuff doorstop. Third, the invention must have a demonstrable function. If you mix silicone with boric oxide and come up with an exceptionally bouncy rubber, you won't necessarily get a patent. Demonstrate its value as a toy, however, and you can call it Silly Putty and make a fortune. The final guideline requires "enablement": The invention must be described clearly enough in writing so that any skilled practitioner in the same trade can read it and fashion the invention himself. A patent does *not* confer ownership. It is simply a contract between the inventor and the government, whereby the former agrees to make public his invention in exchange

40 James Shreeve, *The Genome War*, p. 259.
41 Ibid., p. 270.
42 Ibid., p. 44.
43 Ibid., p. 83, emphasis added.
44 Ibid., p. 109.

> for legal protection against others making or using it for commercial purposes for the next twenty years....
>
> ....Under the first guideline, an organism or any part of one in its natural living state is clearly unpatentable, since it does not originate with the inventor. But in 1972, Herbert Boyer and Stanley Cohen won a patent on the *process* they had invented to manufacture human insulin by cloning its gene. That same year, microbiologist Ananda Chakrabarry applied for a patent on a microbe he had constructed that could degrade crude oil. The utility of his invention was obvious, as was its nonobviousness, and Chakrabarry had no problem writing down the recipe so that any other skilled biologist could produce the microbe. But the patent examiner disallowed the application, arguing that microorganisms are products of nature and therefore not original. Chakrabarry appealed the decision, and by 1980 the case had found its way to the U.S. Supreme Court. In a landmark decision, the court ruled in his favor, on the ground that "anything under the sun that is made by the hand of man" is patentable subject matter, including the specialized life form he had engineered into being. Mother Nature may have supplied the ingredients, but Chakrabarry baked the cake.[45]

As will be discovered later in this chapter, these legal issues raise profound implications for the understanding and interpretation of some very old texts.

C. The Potentialities of Genetic Engineering

Chakrabarry's genetically engineered microorganism reveals the full potentiality — both for good and ill — of genetic engineering and, moreover, the private "ownership" or use of such engineered creations. One may envision genetically engineered organisms that literally eat nuclear waste and discharge it as harmless waste,[46] or genetically engineered drugs,[47] free of the harmful effects of ordinary pharmaceuticals. But by the very same token, one may imagine a world of horrifying possibilities, of genetically tailored plagues, diseases, and viruses that target only races or people with specific genetic characteristics, or a world of horrible chimerical creatures, part human, part animal, engineered for a specific purpose, or devoid of normal human compunction and compassion, literal genetic "Manchurian Candidates" that, coupled with the technologies examined in chapter three, would be the perfect killing

45 James Shreeve, *The Genome War,* pp. 227–228, emphasis in the original.

46 James Shreeve, *The Genome War,* p. 112.

47 Ibid. p. 121.

machines. Strangely, it is in the engineering of chimeras — of genetic *hybrids* — that we find the clearest link to ancient times and texts, and to some very disturbing potentialities and scenarios.

D. The "Mesopotamian" Genome War: The O'Briens Again

The Babylonian "creation epic" *Enuma Elish* — which I have elsewhere interpreted as a "war epic" and not a creation epic at all[48] — contains the slightest hint of the possibility of genetic engineering of chimerical beings in ancient times, referring to "scorpion men" and "fish men."[49] Of course, these ambiguous references could mean almost anything, from metaphorical expressions to fully-fledged expressions of the role of a genetic technology in that war.

But the *Enuma Elish* is hardly the only ancient Mesopotamian text containing statements suggestive of an active technology of genetic engineering in play in paleoancient[50] times. Indeed, in the *Atra-Hasis* epic, the creation of mankind himself is described both in grisly terms and in terms that strongly suggest that mankind is himself one such chimerical creature, a genetic mixture of "the gods" and of some pre-existing terrestrial hominid. Here is how the academic translation of Sumerologist Stephanie Dalley recounts the story:

> When the gods instead of man
> Did the work, bore the loads,
> The gods' load was too great,
> The work too hard, the trouble too much.
> The great Anunnaki made the Igigi[51]
> Carry the workload sevenfold.
> Anu their father was king,
> Their counselor warrior Ellil,
> Their chamberlain was Ninurta,
> Their canal-controller Ennugi,
> They took the box [of lots]....,
> Cast the lots; the gods made the division.
> Anu went up to the sky,

48 Farrell, *The Cosmic War: Interplanetary Warfare, Modern Physics, and Ancient Texts* (Kempton, Illinois: Adventures Unlimited Press, 2007), pp. 150–162.

49 Ibid., p. 153.

50 The somewhat redundant term "paleoancient" is my term to designate the possibility of a pre-existing Very High Civilization that preceded the known civilizations of antiquity, Sumer and Egypt.

51 "Igigi," i.e., man.

[And Ellil(?)] took the earth for his people (?)...[52]

They were counting the years of loads.
For 3,600 years they bore the excess,
Hard work night and day.
They groaned and blamed each other,
Grumbled over the masses of excavated soil;
"Let us confront our [ ] the chamberlain,
And get him to relieve us of our hard work!
Come, let us carry [the Lord(?)].
The counselor of gods, the warrior, from his dwelling..."[53]

As I observed in my book *The Cosmic War,* these passages make it clear that the "gods" were very near "open revolt due to the exorbitant workload laid on them, and they demand to see the 'chamberlain.'"[54]

A little later in the epic, the "strike" threatens to become an open revolt or civil war:

"Every single one of us gods declared war!
We have put [a stop] to the digging.
The load is excessive, it is killing us!
Our work is too hard, the trouble too much!
So every single one of us gods
Has agreed to complain to Ellil."[55]

Seeking some way to ease their burden, the "gods" decide to create an intelligent worker for the precise purpose of being a worker, a serf, a slave, a function that we shall say more about toward the end of this chapter:

Ea[56] made his voice heard
And spoke to the gods his brothers...
...
"There is [ ]
Belet-ili the womb goddess is present —
Let her create primeval man

52 Stephanie Dalley, *Myths from Mesopotamia,* p. 9.

53 Dalley, *Myths from Mesopotamia,* p. 10.

54 Farrell, *The Cosmic War,* p. 141.

55 Dalley, *Myths from Mesopotamia,* p. 12. Ellil is simply another form of the name of the god Enlil.

56 "Ea," another form of the name of the god Enki.

> So that he may bear the yoke [( )],
> So that he may bear the yoke, [the work of Ellil],
> Let man bear the load of the gods!"[57]

Then, further on in the text, the rather grisly methods of the creation of mankind are described in detail:

> Enki made his voice heard,
> And spoke to the great gods,
> "On the first, seventh, and fifteenth of the month
> I shall make a purification by washing.
> Then one god should be slaughtered.
> And the gods can be purified by immersion.
> Nintu shall mix clay
> With his flesh and blood.
> Then a god and a man
> Will be mixed together in clay.
> Let us hear the drumbeat forever after,
> Let a ghost come into existence from the god's flesh,
> Let her proclaim it as his living sign,
> And let the ghost exist so as not to forget (the slain god)."
> They answered "yes!" in the assembly,
> The great Anunnaki who assign the fates.[58]

After taking the decision to create mankind, a serf-worker, the Anunnaki "gods" "then proceed to the task of slaughtering one of their own and creating 'primeval man.'"[59]

> In the first, seventh, and fifteenth of the month
> He made a purification by washing.
> Ilawela who had intelligence,
> They slaughtered in their assembly.
> Nintu mixed clay
> With his flesh and blood.
> They heard the drumbeat forever after.[60]

As I observed in *The Cosmic War,* this account — a more or less standard type

57 Dalley, *Myths from Mesopotamia,* p. 14.
58 Dalley, *Myths from Mesopotamia,* p. 15.
59 Farrell, *The Cosmic War,* p. 142.
60 Dalley, *Myths from Mesopotamia,* p. 15.

of academic translation of the text — gives a "crucial insight into the 'morality' of the Anunnaki, who are clearly not above murdering one of their own to lighten the workload of the rest" in order to create their worker-serf, man.[61]

Enter Christian and Barbara Joy O'Brien and their landmark book *The Genius of the Few* once again, for they maintain that such standard exercises of academic translation may be missing some significant clues. But in order to appreciate the case that they argue, it is necessary to understand it in the light of their wider methodological assumptions, for these in turn raise, once again, issues for religious apologetics.

The O'Briens state the theme and sources consulted in their book in the following fashion:

> *The Genius of the Few* is an account of the activities of a group of culturally and technically advanced people who settled in a mountain valley in the Near East around 8200 B.C. and, as their primary concern, established an agricultural centre for the teaching and training of local tribesmen. Their secondary activities were even more dramatic if the accounts which we have from Akkadian sources, and our interpretations of them, are to be accepted.
>
> The records of these Shining Ones, as we prefer to call them, are taken from three principal sources: (a) Sumerian tablets from the Library of Nippur on which they are referrred to as the (Anunnaki); (b) ancient documents from the Hebraic Books of Enoch where they are described as Angels; and (c) a critical interpretation of the biblical Book of Genesis which uses the Hebrew words *ha elohim.* [62]

It is with the first of their sources — the Akkadian tablets from the Library of Nippur — and with their interpretation of them that we will be chiefly concerned here.

Additionally, the O'Briens believe that archaeology and, by implication, paleography itself, are in part to blame for what they believe is a massive misinterpretation — and, by implication, mistranslation — of these ancient stories:

> In this sphere, one small, undetected error in comparison can become self-perpetuating and mar the whole fabric of the interpretation. And even worse, one small, but attractive, error in interpretation can set up a chain reaction that can lead to whole histories being wrongly conceived.[63]

61 Farrell, *The Cosmic War,* p. 142.

62 Christian and Barbara Joy O'Brien, *The Genius of the Few,* p. 17, emphasis in the original.

63 Christian and Barbara Joy O'Brien, *The Genius of the Few,* p. 20.

For the O'Briens, this background forms the basis for three significant methodological assumptions that they make, which informs the whole of their book. Christian O'Brien states these three assumptions with astonishing clarity:

> ...(Archaeologists) certainly did not have the answer to the development of the Middle East — **because behind each successive magnificent advance there had lain, undetected, the arcane stimulus of the genius of the few.** Or had it been detected, and not recognized? I found that I needed fresh skills in the study of this arcane influence. I turned to languages — Sumerian, Hebrew, Greek, and even Gaelic where it was required; I translated tablets from Nippur that had not been touched in almost a hundred years, and unearthed Hebrew books that, until a few years ago, had not been available for study for more than fifteen centuries. Gradually, the doubts crystallized — I was not sure that we were right — **but I was certain as I was of my own existence that scholars, over five millenia [sic], had gone very wrong.**
>
> And, in taking the path that they had, they had established concepts that never should have been in the receptive minds of men. **Fundamentally, early scholars had given too little weight to those deification processes to which the later Sumerians, and the Babylonians, were fanatically prone... making gods where none had existed...** and hiding glories of erudition and altruistic activity behind the mask of the shrine. **Moreover, later archaeologists, and anthropologists, too, followed in that same path which tends to confuse the secular with the religious; which turns palaces into temples, houses into shrines, customs into rites; and makes every buried statuette a religious relic.**
>
> Out of the earliest of these unfortunate mistakes, there grew a strange religious tradition that fed, avidly, upon itself, and grew stronger with every act of worship and ritual repetition. Shining-countenanced Lords of Cultivation, as they were described in ancient writings, became blurred and distant memories... and were elevated to gods. And the leaders of those same Lords became Gods; and the supreme commander of them all — Great *Anu* to the Sumerians, and the *Most High* to the Hebrew — vicariously became GOD.
>
> And all that time, the true God, the Spirit who is the ultimate arbiter of all Mankind, remained unknown to all but the mystics... as far above those resplendent creatures as we are about the worms in the field.[64]

64 Christian and Barbara Joy O'Brien, *The Genius of the Few,* p. 24. All ellipses are in the

Note carefully what the implications of Christian O'Brien's statements really are for the methodology by which he proposes to interpret the texts he translates and examines:

1) He presupposes the existence of a *hidden elite* that is not of the species *Homo sapiens sapiens* quietly guiding human civilization;
2) He presupposes that the fundamental error of academic examination of the artifacts and texts of the earliest period from this region is to interpret everything as being an example of the deification processes of the Sumerians and Akkadians themselves, whereas, he maintains, if examined from a wholly *secular* point of view, a radically different picture emerges of what those texts might actually be saying;

But what of the *third* methodological assumption? O'Brien notes that one primary feature of these ancient languages is *paronomasia:*

> Another source of ambiguity lies in the fact that early Middle Eastern languages leant heavily on *paronomasia* to give variety to simple phrases — **a form of punning which allowed several different meanings to be given to a single set of symbols.**[65]

Thus we have the third component:

3) The paronomastic nature of ancient languages of the region permits one to translate such texts by a wholly different paradigm — the radically secular one — by noting the plain literal meaning of words, and allowing for the possibility that a primitive language is describing advanced technology in use. The error of academic translations is not, therefore, in mistranslation of individual words or phrases, for given the paranomastic nature of those languages, such interpretations are plausible. But, says O'Brien, they are not the *only* possible ones, and the implication of his remarks is that if there is an error in those translations, it is a *paradigm error* and *not* a *philological* one. Or, if one wishes to speculate a bit more, the paranomastic nature of these languages permits the true nature of the history being recounted to be hidden behind a religious patina, a "psychological opera-

original and thus no text is omitted in the quotation. Boldface emphases added.

65 Christian and Barbara Joy O'Brien, *The Genius of the Few,* p. 25, boldface emphasis added. The O'Briens make the truly astonishing assertion that "before the advent of the Semitic influence, the Sumerians had no shrines, and built no temples." (p. 37)

> tion" called religion, which only served to empower the same hidden elite. This, fundamentally, is the general methodological principle of the O'Briens, as is evident from their remarks cited above.[66]

This brings us to the central text in their "paradigm reinterpretation," the so-called Kharsag Epic, or Kharsag Tablets, a set of tablets first published and translated by George Aaron Barton in 1918.[67] While standard scholarship tends to view these tablets as more or less disjunct from each other, the O'Briens believe they form a more or less contiguous narrative.[68]

1. The Anunnaki and the Engineering of Man

What, then, emerges from these texts concerning the creation of mankind, if one adopts their "secular paradigm" of interpretation, and their assumption of the existence of a hidden elite? Here the crucial tablet is the 8th tablet. The O'Briens begin by noting that even in the book of Genesis there is a residue of the Sumerian idea that mankind was created to be a worker-serf for the gods, for he was created and placed in the Garden of Eden for the purpose of tilling and keeping it.[69] This is an important point, for it implies that the account in Genesis chapter two is a heavily edited text, since it omits crucial details of the earlier cuneiform version, if indeed the cuneiform formed any sort of basis for the Genesis account.

The O'Briens then cite the relevant portion from the Kharsag tablets detailing the creation of mankind (and note the differences in translational style between them and Stephanie Dalley); Enki has the floor, and is addressing the other "gods":

> "What are we accusing them of? Their work was very heavy, and caused them much distress [... ... ...] while Belet-ili, the creator of life, is present. Let her create a 'lullu' — a man, and let the man do the work, and carry the burden of the toil of the lordlings...
>
> "While Belet-ili,[70] the creator of life, is here, let her create offspring, and when they become men, let them bear the toil of their lordlings."

---

66 Their observations recall my own thoughts on the "unified intention of symbol" and the encoding of scientific data in ancient myths. Q.v. my *The Cosmic War,* pp. 74–83, and *The Giza Death Star Destroyed,* pp. 49–52.

67 George Aaron Barton, *Miscellaneous Babylonian Inscriptions,* New Haven: Yale University Press, 1918. Reprinted by Kessinger Publishing.

68 Christian and Barbara Joy O'Brien, *The Genius of the Few,* pp. 37–41.

69 Ibid., p. 152.

70 The O'Briens place an explanatory footnote here: "Belet-ili = Mistress of the Lords" was Nankharsag, or Ninlil.

They sent for Ma-mi, the creator of life, and told her: "You are the biological expert,[71] the creatress of Mankind, we want you now to create a *lullu* so that he may undertake the tasks assigned by Enlil, and so relieve the toil of the lordlings."

In reply, the Lady of Creation said to the (Anunnaki), "It is not possible, for me to make these things on my own; Enki has the skills I need. As he can purify everything [or everybody], let him prepare the material that I need."

We now reach difficulties in the interpretation. The text continues by Enki proposing to make a purifying bath on three separate days, roughly a week apart in which he wishes all the lordlings to be dipped for cleansing. Then he requires that one lordling be slaughtered, and that Nintu should mix "clay" from his flesh and blood. The verbatim text is as follows.

Enki opened his mouth
And addressed the great gods
"On the first, seventh, and fifteenth day of the month
I will make a purifying bath.
Let one god be slaughtered
So that all the gods may be cleansed in a dipping.
From his flesh and blood
Let Nintu mix clay,
That god and man
May be thoroughly mixed in the clay
[... ... ...].[72]

Thus, even on a standard sort of "academic translation," mankind emerges as two things:

1) a creature deliberately created or *invented* for the specific *function* of being a laborer, a serf, for the "gods"; and,
2) a creature that is a *chimera, a hybrid* of two different creatures: (a) the gods, and (b) some presumably pre-existing terrestrial hominid.

---

71 The O'Briens note that translators prefer the term "birth-goddess," but add that "a more scientific expression is required here"; again, this follows from the secular paradigm by which they translate.

72 Christian and Barbara Joy O'Brien, *The Genius of the Few,* p. 155, translating BM 78 257(G), Column ii, emphasis added.

But it is precisely here, in this passage, that the O'Briens assert that the translational "paradigm error" occurs to obscure what is really taking place:

> As with the translations of the early chapters of Genesis, something has gone wildly wrong! From what has preceded, the reader will appreciate that the great (Anunnaki) were not such ninnies, or such scoundrels, as to murder one of their own people and then require Ninlil to mix "clay" from his flesh and blood. Nor, later, to spit on the mixture in the hope of producing a hybrid from man and lordling! In any case, how does man take part in the hybridization?
>
> We are satisfied that the authors of *Atra-Hasis* have produced the best translation possible from the Akkadian text. The fault must have lain with the Akkadian scribes who misinterpreted the original texts. Now, the question is — can we, with the material supplied, provide a more realistic account?[73]

The fault in this one instance, in other words, was not in the modern scholarly translations, but rather in a mistranslation original to the Akkadian, or, as one scholar of Sumerian grammar aptly quipped, "One may say that we see Sumerian through an Akkadian glass darkly."[74]

So how do the O'Briens apply their more secular approach to uncover what they believe to have been an original scribal mistranslation?

> In the first place, the translation of the term *ri-im-ka* is suspect. The root word *rimku* does, indeed, mean "washing," but need not imply a bath. The word can also mean "pouring out," and in that context could be translated as "draught"; and it is more likely that all the lordlings would be given (blood) purifying draughts on the first, seventh, and fifteenth days of the month before one was chosen for the experiment, than that they would be given weekly baths.
>
> In the second place, it is not necessary to slaughter someone in order to obtain their purified blood. Thirdly, a mixture of flesh and blood does not make clay. But it could make what we, today, would call a culture. And out of the right kind of culture, it is possible to produce a hybrid of two individuals — it is now standard practice in the production of test-tube babies. Moreover, in the text that follows, the "clay" that Nintu mixed was placed into the wombs of foster

73 Ibid., pp. 155–156.

74 Dietz Otto Edzard, *Sumerian Grammar* (Leiden: Brill, 2003), p. 7.

mothers, who in due course produced the hybrid babies. What, then, was this clay that Nintu mixed?

It was something which, when mixed with "spittle," produced a culture which could be put into wombs to grow into embryos. The Akkadian term for spittle was *ru-tu* or *ru-u-tu;* and if this were, originally, loaned from the Sumerian, it could have meant a "conception escape." And an escape of "semen" is almost indistinguishable from "spittle."[75]

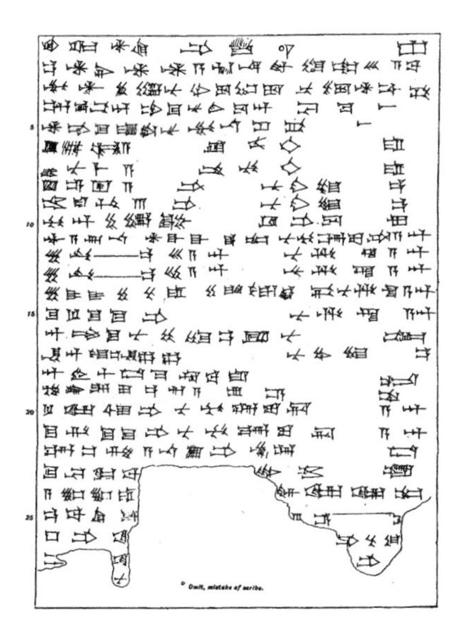

The 8th Kharsag Tablet, Obverse[76]

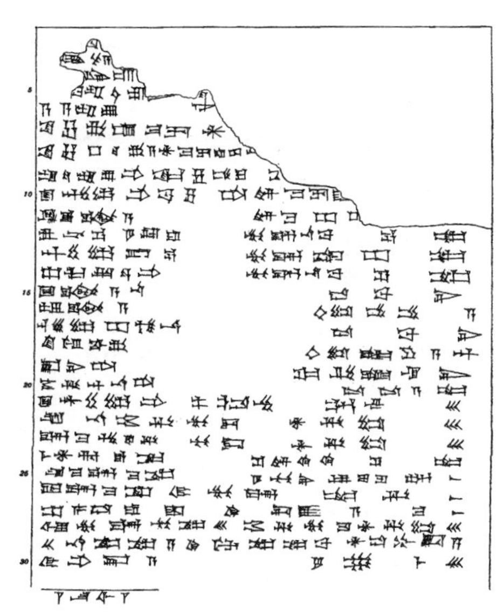

The 8th Kharsag Tablet, Reverse[77]

2. The O'Briens on the Technological Indications

What then of the "clay" or "culture"? Could a similar case be made for a corrupted Sumerian origin for the term?

The Akkadian word for this "clay" was *tittu;* written as *ti-it-tu.* In the context of the hybridization story, the Akkadian word is meaningless to us — and we believe that it was the same to the Akkadian scribes. This suggests that the word was copied from the original Sumerian

75 Christian and Barbara Joy O'Brien, *The Genius of the Few,* p. 156.

76 George Aaron Barton, *Miscellaneous Babylonian Inscriptions,* plate xviii.

77 Barton, *Miscellaneous Babylonian Inscriptions,* plate xix.

without alteration — and we know that the original was in Sumerian, because the personalities mentioned, like Enlil, Enki, Nusku and Anu, were from the Sumerian pantheon rather than the Semitic one. Consequently, we must ask ourselves what *ti-it-tu* could have meant to the Sumerians.

The syllables had the following archaic signs and meanings:

At this juncture, the O'Briens reproduce the following table:

ti	'to live', 'live', or 'living'.
or *ti(n)*	'life' — probably an early variant of the above.
it or *id*	'with'.
tu	'bear', 'beget', 'enter'.
or *tu(n)*	'Portion', 'piece', or 'increase'.

They then comment as follows:

> All these meanings appear to be apposite to the subject matter so, consequently, it is possible to consider that the "clay" which Nintu was to mix with lordling sperm (spittle) was a "piece of life" or "material of life." The jump that must be taken, here, is to conclude that this "piece of life" was a "female ovum." And this jump is justified by the closing stages of the narrative where fourteen "mothers" are brought into the story to carry the resulting embryos.[78]

With this secular, "technological" paradigm of interpretation in hand, they note that "it should be possible to continue the narrative with some degree of credibility."[79] The translation resulting from this paradigm shift is rather intriguing for its implications:

> Enki said: "On the first, seventh, and fifteenth day of the month, I will prepare purifying draughts; let all the lordlings be purified, and then choose one from whom to take a blood sample. From his flesh and blood we will take what is required for the Lady of Creation to mix the life cultures, so that the lordling and man may be thoroughly hybridized in the culture. *Let the hybrid receive "spirit" from the lordling's flesh, and then, let us not forget that living man will have a "spirit."*

78 Christian and Barbara Joy O'Brien, *The Genius of the Few,* pp. 156–157.

79 Christian and Barbara Joy O'Brien, *The Genius of the Few,* p. 157.

> In the Assembly, the Great (Anunnaki), who administer the affairs of Earth, agreed. On the first, seventh, and fifteenth day of the month, Enki made the purifying draughts. We-ila, who had the right characteristics, was chosen by the Assembly to give blood. And from his flesh and blood, the Lady of Creation mixed the cultures.
> [... ... ...]
> After she had mixed the cultures, she summoned the (Anunnaki), the great lords.
> The lordlings [the great lords] gave their sperm for the life cultures.[80]

Thus, on any reading, if the text is to be credited with any veracity — a "standard academic" translation such as Dalley's, or in a more secularized translation such as the O'Briens' — mankind is an engineered hybridized creature, part "lordling" and part "human." In the O'Briens' view, the sperm donor came from a "lordling," or one of the Anunnaki, and the egg donors were human.[81]

But what of the assertion that one of the "gods" had to be slaughtered in order to bring this to pass? Again, the paradigm shift and methodological assumptions lead the O'Briens to a very different translational conclusion:

> ...(It) is far more probable that the (Anunnaki) wished, not to kill, but to "shed blood" for the purpose of obtaining blood samples from which they could select the most suitable of the lordlings.
>
> The importance of the blood sample in determining histocompatibility is well known in modern medicine, because foetal wastage due to blood-group incompatibility forms a serious proportion of stillbirths. It is known, for example, that O-group mothers will more often carry incompatible foetuses than mothers of other blood-groups; in fact, where B- and AB-group males marry O-group females, such matings are termed ABO-incompatible. If this occurs so frequently among modern, polymorphous, but related populations, it may have been far more serious a problem when attempting a conjugation of (Anunnaki) and Hominid, and may have required a very careful selection of blood type.
>
> Even among apparently compatible types, it may have been necessary to make a detailed study of the effect of important antigens on suitable cells and tissues, notably on both blood cells and skin cells, to determine their effect on histocompatibility. This may well be the explanation for the use of both blood and "flesh" in the

80 Ibid., emphasis added.

81 Ibid., p, 158.

experiments carried out by Enki and Ma-mi. In fact, the purification process mentioned may have referred to the process of neutralizing incompatible antigens in the male donor.

The Lady of Creation so manipulated the cultures that lordling and man were "thoroughly mixed." In genetic terms, this mixture was to be half Lordling and half Human; and since the former are stated to have provided the male elements, the female elements must have been taken from human women; and these women could only have come from the Cro-Magnon tribes in the vicinity. Earlier in the epic, Enki is quoted as saying: "You are the biological expert, the creatress of Mankind, we want you now to create a *lullu...*" From this we might infer that, in her role of biologist to the (Anunnaki), Ninlil had previously been active in creating Mankind, and this *lullu* was to be an *ad hoc* operation, possibly resulting in a specialized hybrid, bred for heavy labour.

Such an inference would raise problems. Mankind, at that time, was the Cro-Magnon race and admittedly, it had appeared very suddenly, and markedly superior to its contemporaries — the Neanderthalers. But that event was thirty thousand years earlier; and it would have had to have been another biological expert from another (Anunnaki) group. Could the (Anunnaki) have descended twice onto the Near East? And could they have carried out an earlier hybridization between themselves and the Neanderthalers — to produce the remarkable advance from Neanderthal to Cro-Magnon?

In answer to these two imponderable questions, we can only state that there is no evidence that they did so; but absence of evidence cannot be taken as evidence of absence. They might have![82]

Their technological "translation" of texts, however, reaches its true zenith when they turn to other considerations in the text.

The implementation of the hybridization project took place in the *Bit Shimti. Bit* was the Akkadian for "house" but *Shimti,* or *si-im-ti,* is a word that might have the seed of surprise within it. The archaic pictograms of the Sumerian syllables may be analyzed as follows...

They then reproduce the following insert:

82 Christian and Barbara Joy O'Brien, *The Genius of the Few,* p. 161.

(i)						
	= *igi*	=	INU	=	'eye'
[cuneiform sign]	= *igi*	=	NAMAR	=	'be bright'
	= *si*	=	SI	=	'see, look'

O'Briens' Sumerian Insert[83]

The O'Briens then suggest that the meaning is simply what is literally suggested: a "bright eye for seeing."[84]

Then they follow this up with another insert of the Sumerian cuneiograms of the phonetic syllables:

(ii)	[cuneiform sign]	= *imi*	= TITTU	= 'clay' (life culture)*
(iii)	[cuneiform sign]	= *ti*	= EMU	= 'observe' or 'examine'

O'Briens' Sumerian Cuneiogram Insert[85]

It is best to cite the O'Briens directly on what they believe is the technology signified by this odd assortment of word roots: "The best combination of these meanings is 'bright eye for examining the life culture.' The Bit Shimti may well have been the building which housed this piece of apparatus which appears to indicate an illuminated miscroscope."[86] That is possible, but given all the preceding discussion of the modern genome project, and the vast array of equipment needed to sequence the genome much less *splice* the genes of one organism to another, the meaning might equally be taken to indicate the sequencers and color-coded light-readers themselves. In short, to make such a reading as the O'Briens' work, much more technological resources would be needed than mere microscopes. This indeed might point to a fundamental difficulty with their "paradigm shift in translation," for the sheer *size* of a project to do what they are suggesting is implied by the scope of the task itself, and yet the texts that they cite do not seem to hint of a project of such scale. Moreover, as we have seen in the first parts of this chapter, such a project would require massive computational power, and again, there are no hints of this in the texts they cite. Regrettably, the O'Briens do not make a complete translation of all the Kharsag tablets in their book, which might yield such information.[87]

83 Ibid., p. 162.

84 Ibid.

85 Christian and Barbara Joy O'Brien, *The Genius of the Few*, p. 162.

86 Ibid.

87 Again, the tablets are reproduced in Barton's *Miscellaneous Babylonian Texts*.

What does all this add up to in the O'Briens' opinion? There are, they maintain, but two possibilities for the genetic constitution of mankind, if one takes these ancient tablets — and their own interpretation of them — seriously:

> 1. If Cro-Magnon Man were a hybrid of Neanderthal Man and (Anunnaki), and the Patriarchal tribes were hybrids of Cro-Magnon Man and the (Anunnaki), then the Patriarchal tribes-people — who were the progenitors of the Jewish race — were three parts (Anunnaki) and only one part Neanderthal.[88]
>
> 2. Alternatively, if Cro-Magnon were not a hybrid, but an evolutionary mutation of Early Man, then the Patriarchal tribes were half (Anunnaki) and half Early Man.
>
> In either case, it has to be stated that the Jewish race, through their Patriarchal progenitors, carry more of the "divine" (Anunnaki) strain within their cells than us Gentiles. The percentages would be roughly as follows:

	Patriarchal Tribes	**Gentiles**
Case 1:	**75% (Anunnaki)/25% Hominid**	**50% (Anunnaki)/50% Hominid**
Case 2:	**50% (Anunnaki)/50% Hominid**	**Nil (Anunnaki)/100% Hominid**

It is worth citing how the O'Briens continue to explore this question.

> Which case is correct? There are only four clues — the first lies with the infant Noah who was so startlingly like the Shining Ones, and so startlingly unlike his own family, that his father was constrained to beg Methuselah to make the forbidden journey to Eden to obtain reassurance from Enoch.[89]

They then cite the Book of Enoch, 106: 1–8:

> After some time, my son Methuselah took a wife for his son, Lamech, and she became pregnant by him and gave birth to a son. The child's

88 Note that this first explanation begs the question that they themselves earlier insisted had no evidence, namely an *earlier* hybridization project leading to the creation of Cro-Magnon man.

89 Christian and Barbara Joy O'Brien, *The Genius of the Few,* pp. 162–163.

> body was as white as snow and as red as the rose, and the hair of his head was in long locks which were as white as wool: and his eyes were beautiful. When he opened his eyes, he lighted up the whole house like the Sun might have done; the whole house was bright. And he straightaway sat up in the hands of the midwife, opened his mouth, and spake of the Lord of Justice. His father, Lamech, was afraid of him, and ran to his father, Methusaleh.
>
> And he said to him: "I have produced a strange son, different from, and unlike Man; he resembles the Sons of the Lord in Eden. His nature is different, he is not like you and me — his eyes are like the rays of the Sun and his face shines. It seems to me that he is not born of my stock, but that of the Angels..."[90]

Their comment on this strange development is also worth citing:

> It would appear that the young Noah was a throwback to his progenitor on the male line — We-ila. Of course, this would have been more likely to happen under Case 1, but could still have happened under Case 2.[91]

Or to put it in rather more stark terms — ones that reveal the complete implications — Noah was himself a product of the *b'nai elohim* or "sons of God" (or "sons of the gods") of Genesis 6, and of the daughters of men: *Noah himself was a hybrid,* and suggestively, the implication is that he might not really have been the biological son of Lamech at all.

For the O'Briens, then, there is a hidden reason why Noah and his family were "saved" from the flood, for he clearly bore a "divine" bloodline that "the powers that be" or rather "the powers that were" wished to preserve.[92]

But it is the "fourth clue" in which the O'Briens find the "definitive key" to unraveling the mystery, for some of the Anunnaki, the fallen ones designated "Watchers" in the Book of Enoch,

> ...were able to procreate with the daughters of the Patriarchs. For this to have been possible, in a natural situation outside of the laboratory, there must have been a very close genetic relationship between Watcher and Woman. This would have been favoured

---

90 Christian and Barbara Joy O'Brien, *The Genius of the Few,* p. 163. I cite the O'Briens' translation because, in their view, the Bright or Shining countenance was a distinguishing physiological feature of the Anunnaki.

91 Ibid.

92 Christian and Barbara Joy O'Brien, *The Genius of the Few,* p. 163.

> by the higher (Anunnaki) genetic ratio in the Patriarchal woman assumed in Case 1.[93]

In other words, the O'Briens incline to the view that there *was* an earlier project to hybridize Cro-Magnon man, and that from this experiment yet another hybridization occurred between the "gods" and that Cro-Magnon man to produce the hybrid of a hybrid, modern *Homo sapiens sapiens.* In chapters seven and eight, we shall see just how closely this harmonizes not only with other ancient texts and legends, but with aspects of the standard model of human origins within science as well.

A hint of it is given in the Mayan creation epic, the *Popul Vuh*, an ocean away in Meso-America:

> The *Popul Vuh* states that mankind had been created to be a servant of the "gods." The "gods" are quoted:
>
> "*Let us make **him** who shall nourish and sustain us! What shall we do to be invoked, in order to be remembered on earth? We have already tried with our first creations, our first creatures; but we could not make them praise and venerate us. So, then, let us try to make obedient, respectful beings who will nourish and sustain us.*"
>
> According to the *Popul Vuh*, the "gods" had made creatures known as "figures of wood" before creating *Homo sapiens*. Said to look and talk like men, these odd creatures of wood "existed and multiplied; they had daughters, they had sons..." There were, however, inadequate servants for the "gods." To explain why, the *Popul Vuh* expresses a sophisticated spiritual truth not found in Christianity, but which is found in earlier Mesopotamian writings. The "figures of wood" did not have souls, relates the *Popul Vuh*, and so they walked on all fours "aimlessly." In other words, without souls (spiritual beings) to animate the bodies, the "gods" found that they had created living creatures which could biologically reproduce, but which lacked the intelligence to have goals or direction.
>
> ...
>
> Creating *Homo sapiens* did not end Custodial headaches, however. According to the *Popul Vuh*, the first *Homo sapiens* were *too* intelligent and had *too* many abilities!

93 Ibid., p. 164.

> *"They (first* Homo sapiens) *were endowed with intelligence; they saw and instantly they could see far, they succeeded in seeing, they succeeded in knowing all that there is in the world. When they looked, instantly they saw all around them, and they contemplated in turn the arch of heaven and the round face of the earth.*
>
> ...
>
> *"But the Creator and the Maker did not hear this with pleasure. 'It is not well that our creatures, our works, say: "they know all, the large and the small,"' they said."*[94]

The situation is remarkably parallel to the Mesopotamian texts, for there too, after the chimerical mankind — part "god," part "hominid" — is created, the gods quickly complain that the new being is *too* smart. Note also that the reasons given for mankind's creation in the *Popul Vuh* are identical to the reasons given in texts from the Middle East, both Mesopotamian and biblical: the "gods" needed slaves to do their work for them *and to worship them.*

3. Verification: Genetics, Space, and Skeletons

But if all this is true, how might one go about *verifying* it? There are some clues from various texts that afford a speculative basis by which the O'Briens' interpretations of ancient texts and arguments might be verified. The Book of Genesis and other books of the Bible, for example, record the offspring of such unions as being of gigantic or larger than normal stature. Indeed, the word *lugal* itself, often translated as "king" in academic translations, simply means "big man," and this fact, coupled with the large stature of ancient kings depicted in Mesopotamian cylinder seals *might* indicate not a metaphorical and artistic device, but an actual truth.

But something more is required for verification than this.

It is here that one must connect the dots in an unusual fashion. Space anomalies researcher Richard C. Hoagland has demonstrated, in a series of remarkable papers on his website, that the Saturnian moon Iapetus and the Martian moon Phobos show distinctive and persuasive signs of being great artificial bodies.[95] These large bodies, if artificial — and this author believes that Hoagland's case is very strong — would possibly be the natural products of intelligent human-like creatures of large stature. It is thus possible that in addition to the recovery of ancient technology — should humanity ever journey to and explore those

94 William Bramley, *The Gods of Eden* (Avon: 1990), pp. 176–177, emphasis Bramley's.

95 See the papers by Richard C. Hoagland, "A Moon with a View" and "For the World is Hollow, and I Have Touched the Sky," www.enterprisemission.com.

"moons" — we might also discover that *bodies* that could, upon genetic examination, indicate a genetic relationship to humanity.

Additionally, the statements of ancient texts mentioned previously would also seem to imply, here on earth, the existence *somewhere* of the remains of such creatures. Of course, the Internet is rife with "photographs" of such alleged giant remains, and in all cases known to this author, these are elaborate hoaxes. Nonetheless, as we shall discover in the next chapter, there are indications that these remains have been discovered... and quietly spirited away.

4. Whose Agenda, Public, or Private?

As was seen previously, the modern genome race raised profound questions and implications for ethics and jurisprudence: were specific genes, or the processes used to map them, patentable? More importantly, were chimerical, or hybrid life forms, patentable? The last question assumes even more importance *if* the foregoing Sumerian accounts of the origins of mankind are true, for by any translation — whether those of the O'Briens or more standard academic translations — mankind is a chimerical, *engineered* creature.

As such, it is worth recounting what the four requirements for the grant of a patent are under modern American law. To be patentable, an invention or process must:

1) be original;
2) be non-obvious;
3) have a demonstrable function; and,
4) be enabling, i.e., any inventor or engineer should be able to read the patent and be able to reproduce the invention or process it describes.[96]

It is worth recalling that the landmark 1980 Supreme Court decision granted Ananda Chakrabarry a patent on an engineered microbe on the basis that "Mother Nature may have supplied the ingredients, but Chakrabarry baked the cake."[97] So long as "the hand of man" was involved in the engineering, the process — and even the organism itself, if it was not naturally occurring — was patentable.

This places the story of the creation of man as told in the ancient cuneiform tablets into an interesting light, for *if* true, then under the standards of American patent law, the human being as a hybrid creature would apparently

96 James Shreeve, *The Genome War*, p. 227.
97 Ibid., p. 228.

fulfill all four requirements for a patent. As such, human beings, as *chimerical hybrids* of two other species, are (1) original; (2) non-obvious, for they are not the products of nature, but of "the hand of 'man'," (3) were created for a demonstrable function, i.e., were created to be slaves and serfs of the "gods," and (4) they were the result of a process of genetic engineering that was reproducible "by the hand of man," and therefore, the process was "enabling," allowing any competent genetic engineer to reduplicate the process.

These implications compel some speculative questions, for if mankind as currently constituted is a chimerical creature, a genetically engineered creature, and moreover, was created for the express purpose of being a slave to the "gods," then who owns him? Let us speculate: *if* mankind's original owner-creators were suddenly to return — right now — to planet Earth, would they have a legal claim? And would, moreover, they be able to prove it? Would they attempt to re-assert their old hegemony? And what court would have legal jurisdiction to hear such a case? Additionally, one would be faced with two legal claims: (1) that of the returning "owner-creators," and (2) that of the course of performance of humanity since their "departure," which would legally demonstrate humanity's independence and self-governance since their departure. Humanity would, so to speak, be abandoned property and under new ownership, namely, itself. Would these owner-creators attempt to reassert their ownership by demonstrations of force and superior technology, only to discover that humanity can now "shoot back"?

Disturbing questions all, but there is yet another question that looms over them all, and that is, in the cuneiform accounts of the creation of man, is his creation that of a *private* or *"corporate"* entity, or a *public* one? Is the question of "ownership" even relevant? Given all the argumentation of the preceding pages, our inclination is to conclude that mankind was created by a private, corporate entity, and for the service of specific "gods" within the pantheon, for as has been seen, the establishment of a reliably accurate, reproducible standard of weights and measures based on astronomical and geodetic principles, was the work largely of a private elite, and done for the purposes of the ultimate establishment of trade. We must conclude, too, that the status of slavery in early ancient legal codes may be a residue of the creation of mankind himself.[98] All these points argue for a

98 Some might be tempted to argue that mediaeval doctrine of legal atonement and the ransoming of man from the devil are a further residue of these ancient legal conceptions. As of this writing, however, this author can find no direct conceptual or textual link between the ancient texts and the mediaeval doctrine, other than, of course, the texts of the Old Testament law itself. To argue such a case would require a massive reinterpretation of the evolution of the text of the Torah itself, and a presupposition that massive editing is in evidence for the purpose of obscuring external references to Mesopotamian cultures and their influence upon it. Such a task is well outside the scope of this book.

corporate rather than a *state or governmental* elite having been involved in mankind's creation.

Whatever else one may make of the reading of the Kharsag tablets by the O'Briens and others, one thing stands out as an inevitable consequence of such interpretations, and that is, that in order for such a genetic engineering project to have succeeded, the whole modern panoply of scientific discoveries and technologies associated with the modern genome project had to have existed in "paleoancient" times as well: sequencers, microscopes, organic chemistry, the knowledge of the double helix itself, the techniques of splicing, and, of course, a complete "genetic map" of the species involved in the hybridization of man. In short, if the cuneiform tablets are to be believed, then *at least* a similar pitch of genetic science had to exist — if not much greater — in prehistoric times than exists now.

How, then, would one go about establishing the truth, or lack thereof, of the astonishing claims of those tablets? One cannot settle here simply for re-interpretations of those tablets along the lines of the O'Briens, for in that case one is assuming what one is proving. One must have *external* corroboration. And that corroboration, by the nature of the case, can only come from within the modern human genome itself. Is there a "code" within the genetic code that might suggest that we are indeed the creatures of such a project begun and executed long ago? Is there a hint of the "artificiality" of modern man? And if so, who is searching for it, and why? Are there indeed remains of giants that have been quietly spirited away? And if so, why?

Here, as they say, "the plot thickens"...

Six

A Connection of Miscellanies:

The Codes within the Code, and the "Archaeology Conspiracy"

∴

"An amino acid sequence can be encrypted... since the (genetic) code is a homophonic cipher."
—Mathematicians John C. King and Dennis R. Bahler[1]

HUMAN DNA IS WHERE the complex algorithms of computing, where the arcane symbols of higher dimensional physics and mathematics, and where the biology of life and the meta-physics of consciousness all meet in an intricate encrypted minuet. But is there any evidence that all this complexity was the artificial creation of a hidden elite to begin with? Is there evidence, perhaps, of a "code within the code" that might testify to a hidden hand at play in ancient times, one with, perhaps, an agenda?

Oddly enough, there is, and it comes from a very strange and unsuspected place: ancient China, and its Book of Changes, the I Ching. In that, there is another strange connection to Mesopotamia and its ancient myths, and the possibility of yet another agenda in evidence.

A. The Codes Within the Code

It was after the discovery of the double helix and the basic structure of the human genome that some biologists and geneticists noticed a peculiar

1 John C. King and Dennis R. Bahler, "A Framework for the Study of Homophonic Ciphers in Classical Encryption and Genetic Systems," *Cryptologia*, January 1993, Volume XVII, Number 1 (45–54), p. 53.

DNA	I Ching
1. Discovered ten years ago, has existed since life began. All the vital processes of all living creatures whose structure, form and heredity are programmed in precise detail *universal claim.*	1. All processes of living development throughout nature are subject to *one* strictly detailed program (*universal* physical, metaphysical, psychological, moral *claim*)
2. The basis is the plus and minus double helix of DNA.	2. The basis is the manifestation of the world principle in the primal poles yang (__) and yin (_ _)
3. Four letters are available for labeling this double helix: A-T, C-G (adenine, thymine, cytosine, guanine), which are joined in pairs.	3. Four letters suffice for life in all its fullness. 7= resting yang --- 9= moving yang --- 8= resting yin - - 6 = moving yin ®¬
4. Three of these letters at a time form a code word[1] for protein synthesis.	4. Three of these letters at a time form a trigram, a primary image of the eight possible dynamic effects.
5. The "direction" in which the code words are read is strictly determined (®)	5. The "direction" in which the trigram is read is strictly determined ()
6. There are 64 of these triplets known whose property and informative "power" have been explored. One or more triplets program the structure of one of the possible 22 amino acids; quite specific sequences of such triplets program the form and structure of all living creatures, from the amoeba to the iridescent peacock's feather.	6. There are 64 double trigrams precisely designated and described by Fu-Hsi (3000 B.C.) in very vivid and precise images of highly specific dynamic states (e.g., "breakthrough" or "oppression") with in each case six possible variations of this state and subsequent transformation into another one of the 64 hexagrams — a program of fate, as it were, in which each individual is at all times placed to operate the "switch" of fate, from which point onwards the "train" continues along its appointed "line."
7. Two of these triplets have names: "beginning" and "end." They mark the beginning and end of a code sentence of some length.	7. Two hexagrams have names: *before* completion and *after* completion (frequently opening and closing "melodies of fate" in the oracle).

Schönberger's Table of DNA – I Ching Comparisons*

* Dr. Martin Schönberger, *The I Ching and the Genetic Code: The Hidden Key to Life* (Santa Fe, New Mexico: Aurora Press, 1992), pp. 31–33.

thing, indeed, a thing *so* peculiar that it defied all the laws of coincidence and chance: the strong resemblance of the structure of DNA to the structure of the ancient Chinese system of divination, the I Ching.[2]

The I Ching was discovered by Fu-Hsi in 3000 B.C. "through an insight into nature that defies our understanding."[3] Indeed, it is because of this defiance of our understanding that much more may be at work than the accidental, serendipitous discovery of an effective tool of divination, for as we shall see, there may very well be a hidden Sumerian connection to the I Ching.

However, in order to appreciate this astounding connection, we must have a basic understanding of the structure of the I Ching itself and lay it side by side with the basic structure of DNA.

Like DNA's codons with their tripartite structure of information,[4] the I Ching's system of 64 symbols is built of a simple structure called trigrams, which are combinations of three lines. The lines may be solid, or broken. There are eight such trigrams, and in the I Ching, these images are, in the words of the German I Ching scholar Wilhelm, "images of all that happens in heaven and on earth," and as such, are representation of the tendencies of beings in movement. They represent "not objective entities, but functions."[5] The I Ching functions, in other words, in a manner very similar to DNA, for both describe the possible actions of a given person, given a certain set of programmed instructions inherent in the two systems.[6] Given all these similarities between the I Ching and DNA, geneticist Gunther S. Stent wrote that "the congruence between it and the genetic code is nothing short of amazing."[7] Indeed, it is so amazing that the chances of it being a statistical fluke are very small. The I Ching may, in fact, be viewed as the "other half" of DNA.[8]

2 See the remarks of biologist Gunther S. Stent in Dr. Martin Schönberger, *The I Ching and the Genetic Code: the Hidden Key to Life* (Santa Fe, New Mexico: Aurora Press, 1992), p. 147.

3 Dr. Martin Schönberger, *The I Ching and the Genetic Code: The Hidden Key to Life* (Santa Fe, New Mexico: Aurora Press, 1992), p. 61.

4 Ibid., p. 10.

5 Dr. Martin Schönberger, *The I Ching and the Genetic Code: The Hidden Key to Life* (Santa Fe, New Mexico: Aurora Press, 1992), p. 48. It should be noted that within the system of the I Ching, the system begins with a kind of "primordial maleness" or "yang" (q.v. p. 46), representing heaven. Given this primordial masculine unity, the I Ching itself, in yet another parallel with modern scientific views, represents a system based on a broken symmetry in the physical medium (q.v. p. 45), from which a base dipole — yang and yin, a subsequent division of the feminine part of the masculine — emerges.

6 Schönberg remarks, "Is it not possible that in the DNA code — a *script* with significance, meaning, expression, information, impulse and life — we can see and recognize firsthand both spiritual and material structures erupting into the *hyle,* the material?" (Dr. Martin Schönberger, *The I Ching and the Genetic Code: the Hidden Key to Life* [Santa Fe, New Mexico: Aurora Press, 1992], p. 43.) On this view, the human person is, as it were, *transduced* into material existence from the physical medium, making the person itself a non-local phenomenon. (See also his remarks on p. 113.)

7 Ibid., p. 147.

8 Ibid., p. 9.

This means that someone in ancient times *knew* something, and that the possibility arises that Fu-Hsi did not discover the I Ching, but rather, that the I Ching is a legacy of some lost civilization and its science, for ancient China certainly did not have sophisticated knowledge of the genetic code. Just who did has already been seen from the ancient texts examined in the previous chapters: the Anunnaki.

On this view, the I Ching becomes a "probability calculus,"[9] a kind of "psychic computer"[10] with its own topology of "sequent events"[11] and even a catalogue of those events,[12] implying a kind of universal consciousness of which each person is a component.[13] Indeed, the physical medium not only becomes conscious in a certain sense, but is also — in another dramatic parallel with very modern views in physics — an information-creating medium.[14] This makes the universe a "gigantic organism" based upon a triadic structure underlying all things, and even becomes the basis for tantric sex practices, practices rationalized as a direct participation in the medium itself.[15]

B. Back to the Tablets of Destinies

The strong and more than coincidental nature of the correspondences between the I Ching and DNA — and we have only surveyed but a few — raises the possibility that in looking at the two, we may be in fact looking at the fragments of a lost science that once long ago existed in a higher order unity.[16] On this view, both modern genetics and the I Ching itself would be declined legacies of that higher order unity.

Given what has been said here, we may speculate on some of the properties of that "higher structured order" and its properties: (1) it will first be of an analogical nature; (2) it will *create* information in the medium from a broken symmetry; (3) it will have its own "torsion moment" or rotational characteristics (mirrored in the spiral properties of the DNA helix).

This is to say that the DNA of man interfaces directly with the physical medium or aether in a manner that no other DNA of any other species

9 Ibid., p. 17.

10 Ibid., p. 21.

11 Ibid., p. 60.

12 Dr. Martin Schönberger, *The I Ching and the Genetic Code: The Hidden Key to Life* (Santa Fe, New Mexico: Aurora Press, 1992), p. 71.

13 Ibid., pp. 19, 83.

14 Ibid., p. 34.

15 Ibid., p. 15.

16 This view recalls a statement found in the Cooper-Cantwheel set of MAJIC-12 UFO documents where it is explicitly stated: "The laws of physics and genetics may have a genesis in a higher structured order that [sic] once previously thought." See my *Reich of the Black Sun* (Kempton, Illinois: Adventures Unlimited Press, 2004, p. 295).

does, making man a very powerful creature from a certain point of view... and a *threat,* which might account for the fact that the "gods," at least in the Mesopotamian legends, attempt to exterminate him once he was created. In any case, "Man" is, so to speak, made in the image and likeness of the physical medium.[17]

But what we are clearly looking at here is a higher order of structural unity that was understood by *someone* in times that were already "ancient" before ancient times, and we have already suggested where such knowledge and science came from: the Anunnaki, and hence, there may be some connection between the I Ching and the ancient cultures of Mesopotamia. The question is, what is it?

There is one very obvious candidate: the Tablets of Destinies. These, as I have detailed elsewhere,[18] were objects of power, made from some kind of stone or crystal, which enabled the fantastic technologies of the "gods." Moreover, they were also fantastic objects of power and destruction, so fantastic, in fact, that whole wars were fought to possess them, until finally, some of them were deliberately destroyed to prevent their use for warfare again.[19]

The problem is, no one knows for sure exactly what they were, beyond the vague descriptions of stone-like or crystalline objects in those ancient texts. "Putting all [the] indications" from the ancient Mesopotamian texts together, I decided to follow the clues to draw some speculative conclusions on what they may have been. As I outlined in my book *The Cosmic War*, there were essentially six clues:

1. They were, first of all, *information,* or objects *containing* information, that allowed one to tap into "the power of the universe";
2. They were objects of *technology* that in use or conjunction with the "ekurs" or "mountains of stone" — the ziggurats and pyramids of the ancient world — allowed one to *access* that power, and thus wield a global hegemony by dint of being able to manipulate that power *in a variety of ways;*
3. In the myths... examined, these ME[20] are almost always understood to be weapons. Their theft from Enlil by Anzu (or, earlier, by Tiamat!) thus implies something quite important about the civilization of the ancient "gods," and that is, that the whole idea

17 This is a very broad topic, touching directly upon ancient hermetic and esoteric doctrines of man, most especially that of "man as microcosm," and I hope to explore these topics in a future book.

18 See my *The Cosmic War: Interplanetary Warfare, Modern Physics, and Ancient Texts* (Kempton, Illinois: Adventures Unlimited Press, 2007), pp. 204–273.

19 Ibid., pp. 204–233.

20 ME: one of the words used for the Tablets of Destinies.

of "kingship" resided in the implicit ability and threat to make use of this force.

4. As was also seen... the greatest of the MEs, the "Tablets of Destinies," appear to be able to effect action at a distance.
5. Some MEs are also referred to as ME-LAM, or light-emitting, suggesting a connection between the Tablets of Destinies and light, or electromagnetic radiation. This interpretation is supported by the fact that the texts indicate that after their theft, "radiance" disappeared.
6. And finally, it will be noted that these particular MEs, from their first theft by Tiamat to their subsequent theft by Anzu, and their brief (?) period in the possession of Ninurta, exercise a peculiarly corrupting moral influence on their possessors, gradually sapping their will and consuming them with the obsessive desire to control and possess them...[21]

There is, however, another clue about the Tablets of Destinies, and it is both an important one, and one that moreover is a direct speculative conceptual link to the I Ching:

> ...(The) possessors of the Tablets are the least ambiguous concepts in the whole textual and conceptual complex surrounding the mysterious objects. The texts are clear about who possessed them, and more or less clear about the order of their successive owners. These two things — the *possessors* and the *order of the possessors* — afford the key that will unlock the riddle.[22]

This connection to specific owners led me to speculate that "The Tablets of Destinies appear to have been 'activated' only in physical proximity to their possessors,"[23] — in other words, they may have been activated by the certain specific sequences in the DNA of their possessors, and given the fact that the I Ching is a system of divination with so many odd parallels to the genetic code itself, the connection seems both logical and inevitable: *With the I Ching, we may be looking at, so to speak, the "software" remains of the Tablets of Destinies.*[24]

---

21 Joseph P. Farrell, *The Cosmic War*, p. 236.

22 Ibid., p. 237.

23 Ibid., p. 252.

24 There is a more general context into which these observations might also be placed. Many have commented on the qualitative resemblance between *early* Chinese ideograms and the cuneiograms of Mesopotamia, suggesting some sort of deeper though unknown connection between the two

There is yet another possible connection between the I Ching and the Tablets of Destinies, and it comes directly out of physics. Dr. Martin Schönberg, whose book *The I Ching and the Genetic Code* we have been following here, remarks on a peculiar feature of the I Ching that was not even known or speculated about until the advent of modern quantum and nuclear physics:

> The natural direction of rotation is clockwise and shows the sequence of the year. However, it seems very strange to us to say that, knowing that which is to come, depends on backward movement. We have a very long way to search in Western science before we find the theory of "backward" movement, time reversal, the disappearance and appearance of plus- and minus-charged particles, and the calculation of future states, until there is mention of anticlockwise rotation. In fact, we find it only when we come to the results and theories of atomic physics — and of the DNA double helix. During the intervening thousands of years of natural science, there is no mention of plus and minus, time reversal, retrograde movement, clockwise and anticlockwise movement, conversion of energy into matter — except in the natural philosophy of the I Ching (no gods are needed!)[25]
>
> According to the I Ching doctrine of spatiotemporality, there should therefore also be, unfolding from the seeds of the eight primary trigrams (the "world" reading rightwards), a backward path, contrary to the natural order of events, through which the seeds can be recognized, the past understood and the law-governed development of the future predicted — a path which is open to the wise man through his intuitive insight into the course of nature, in accordance with the primary trigrams and their 64 combinations....[26]

This idea of time reversal was also a crucial speculative conclusion that I had arrived at concerning the Tablets of Destinies. After a careful examination of the implied physics of the texts that referred to the Tablets of Destinies, I concluded that they were components of a sophisticated *phase conjugate mirror* able to manipulate the resonant harmonics of objects that exist in the physical medium itself.

civilizations.

25 Dr. Martin Schönberger, *The I Ching and the Genetic Code: The Hidden Key to Life* (Santa Fe, New Mexico: Aurora Press, 1992), p. 51.

26 Dr. Martin Schönberger, *The I Ching and the Genetic Code: The Hidden Key to Life* (Santa Fe, New Mexico: Aurora Press, 1992), p. 52.

Thus, I concluded:

> Grounded both in the ancient texts and a modern physics interpretation of them, the Tablets of Destinies were most likely crystals of some sort through which light or other electromagnetic energy was beamed, and these crystals contained information. This information was the "holographic" interferometric "grating" or "interference pattern," Bearden's quantum potential "template of action," the scalar signatures, of almost every celestial body considered important in, to, and by the civilization of the "gods." These priceless catalogues were, moreover, compiled relative to our own solar system as their physical frame of reference. Additionally, they included within their catalogue of "gratings" or "templates" the "subtle influences" — as an impressed dynamic — of consciousness. They most likely worked best, therefore, when interfaced — by means now lost and unknown to us — with an intelligent and conscious user.... The Tablets of Destinies were the "software" gratings of phase conjugate mirrors for almost every celestial body that was considered important to them...[27]

As indicated above, it would appear that the I Ching is based on some lost and fragmented component of this science and may be the program by which the consciousness of the users of the Tablets was integrated into the whole system.

But what is a "phase conjugate mirror"? Briefly stated, it is a *real* optical technology that really exists, and that was developed in part as a component of President Ronald Reagan's "Star Wars" or "Strategic Defense Initiative" in the 1980s. To understand what a phase conjugate mirror does, one need only understand some simple optics, and why a phase conjugate mirror is such a revolutionary — and potentially deadly — technology.

Part of the Strategic Defense Initiative consisted of the idea of orbiting high-powered lasers in space to shoot down Russian ICBMs upon detection in the early launch stage of their trajectory. The problem was, as any physicist or engineer knew at the time, that atmospheric distortion of the laser beam would occur, thus diminishing considerably the amount of power delivered to the target, and thus requiring that the laser would have to track and stay locked on the target for a much longer period of time in order to destroy it.

However, engineers came up with a way around the problem: phase conjugate mirrors. The idea was to shine two beams on the target and take a

27 Farrell, *The Cosmic War,* pp. 270–271.

"picture" — quite literally, a hologram — of the interference pattern generated by the atmospheric distortion, and then use that hologram as a grid or template through which to shine a laser beam. By doing so, the effects of the atmospheric distortion were literally removed, as if the target beam was traveling *backward in time* to arrive on the target with increased, rather than decreased, intensity. In fact, such waves from a phase conjugate mirror were oftentimes called "time-reversed waves" in the literature.[28]

The idea of phase conjugation may seem far removed from ancient China and its I Ching, until one also recalls that the ancient Chinese tradition speaks of a deadly "weapon of the gods" called a "Yin-Yang" mirror! The name itself is significant, for ordinarily, the primordial masculine Yang nature occurs first. By thus ordering the terms "Yin-Yang" and not "Yang-Yin," the idea of a time reversal, of a system reversal, is inevitably implied. In other words, a phase conjugate mirror!

So what may we conclude from this all-too-brief excursion into physics and divination?

Clearly, someone in ancient times knew something about the profound connection between the structure of DNA and the physical medium itself, and, in the wake of some primordial "Tower of Babel moment," took steps to preserve a component of that knowledge.

In short, we are, once again, dealing with a hidden elite in possession of sophisticated knowledge, who are attempting to preserve that knowledge until the day that mankind was in a similar high scientific development, and could crack the code.

But there is another agenda at work as well....

C. Giants, and "The Archaeology Conspiracy"

1. The New York Times *Reports the Discovery of Giant Remains*

The subject of the discovery of giant skeletons or bones has long fascinated mankind, and never more so than recently, as the ancient stories of the possibility of humanity as a genetically engineered creature have taken increasing hold on the imagination. Indeed, such tales of giants — the offspring of matings between humans and "the gods" if we are to believe the Bible — have given rise to a panoply of photographic hoaxes that make the rounds of the Internet every so often.

But if these ancient stories are to be subject to any sort of scientific verification, then as the previous chapter averred, the possibility of doing so

28 For a fuller discussion of phase conjugate mirrors and their deadly properties, see my *The Cosmic War*, pp. 124–131; 254–263.

lies in the potential discovery of cadavers on the apparently artificial moons of Mars and Saturn: Phobos and Iapetus. Similarly, such remains might be found on earth. Indeed, not all such reports of giants' remains are Internet hoaxes, for on October 3, 1892, the *New York Times* ran a short article, based on a story reported in the *London Globe,* that such remains had been discovered:

> *Race of Giants in Old Gaul*
> From the *London Globe*
>
> In the year 1890 some human bones of enormous size, double the ordinary in fact, were found in the tumulus of Castelnau, (Hérault,) and have since been carefully examined by Prof. Kiener, who, while admitting that the bones are those of a very tall race, nevertheless finds them abnormal in dimensions and apparently of morbid growth. They undoubtedly open the question of the "giants" of antiquity, but do not furnish sufficient evidence to decide it.

Reports such as this abounded in the late 19th and early 20th centuries.

A Race of Giants in Old Gaul.

From the London Globe.

In the year 1890 some human bones of enormous size, double the ordinary in fact, were found in the tumulus of Castelnau, (Hérault,) and have since been carefully examined by Prof. Kiener, who, while admitting that the bones are those of a very tall race, nevertheless finds them abnormal in dimensions and apparently of morbid growth. They undoubtedly reopen the question of the "giants" of antiquity, but do not furnish sufficient evidence to decide it.

October 3, 1892 *New York Times* Article on Discovery of Giant Remains

The problem is, while such reports and discoveries abounded, there was a deafening silence afterward on what those discoveries portended. In fact, the silence was so deafening that after an initial report, the discoveries were never reported nor commented upon again. This has led some to speculate that there is in fact a kind of "archaeological conspiracy" to cover up such discoveries. The question assumes some significance, for if the ancient texts are to be believed, the offspring of the ancient "gods" and humans were giants, and hence, if there is such a cover-up, then this points to some possible agendas at work, not only an ancient one, but a modern one, and perhaps, the *same* one.

2. Egypt in Arizona

One of those advocating such an "archaeology-gate" is the famous ancient anomalies researcher David Hatcher Childress. The beginning of the story concerns the alleged discovery of ancient Egyptian ruins in — of all places — none other than the Grand Canyon! The find was reported in a lengthy article that appeared in the April 5, 1909 edition of *The Phoenix Gazette*. The article is lengthy, but it is worth citing in full.

> Explorations in Grand Canyon
> Mysteries of Rich Cavern Being Brought to Light
> JORDAN IS ENTHUSED
> *Remarkable Finds Indicate Ancient People Migrated From Orient*
>
> The latest news of the progress of the explorations of what is now regarded by scientists as not only the oldest archaeological discovery in the United States, but one of the most valuable in the world, which was mentioned some time ago in the *Gazette*, was brought to the city by G.E. Kinkaid, the explorer who found this great underground citadel of the Grand Canyon during a trip from Green River, Wyoming, down the Colorado River, in a wooden boat, to Yuma, several months ago.
>
> *According to the story related to the Gazette, the archaeologists of the Smithsonian Institute [sic], which is financing the explorations, have made discoveries which almost conclusively prove that the race which inhabited this mysterious cavern, hewn in solid rock by human hands, was of oriental origin, possibly from Egypt, tracing back to Ramses.* If their theories are borne out by the translation of the tablets engraved with hieroglyphics, the mystery of the prehistoric peoples of North America, their ancient arts, who they were and whence they came, will be solved. Egypt and the Nile, and Arizona and the Colorado will be linked by a historical chain running back to ages which stagger the wildest fancy of the fictionist.

Under the direction of Professor S.A. Jordan, the Smithsonian is now pursuing the most thorough explorations, which will be continued until the last link in the chain is forged. Nearly a mile underground, about 1,480 feet below the surface, the long main passage has been delved into, to find another mammoth chamber from which radiates scores of passageways, like the spokes of a wheel. Several hundred rooms have been discovered, reached by passageways running from the main passage, one of them having been explored for 854 feet and another 634 feet. The recent finds include articles which have never been known as native to this country, and doubtless they had their origin in the orient. War weapons, copper instruments, sharp-edged and hard as steel, indicate the high state of civilization reached by these strange people. So interested have the scientists become that preparations are being made to equip the camp for extensive studies, and the force will be increased to thirty or forty persons.

Before going further into the cavern, better facilities for lighting will have to be installed, for the darkness is dense and quite impenetrable for the average flashlight. In order to avoid being lost, wires are being strung from the entrance to all passageways leading directly to large chambers. How far this cavern extends no one can guess, but it is now the belief of many that what has already been explored is merely the "barracks," to use an American term, for the soldiers, and that far into the underworld will be found the main communal dwellings of the families. The perfect ventilation of the cavern, the steady draught that blows through, indicates that it has another outlet to the surface.

Kinkaid was the first white man born in Idaho and has been an explorer and hunter all his life, *thirty years having been in the service of the Smithsonian.* Even briefly recounted, his history sounds fabulous, almost grotesque.

"First, I would impress that the cavern is nearly inaccessible. The entrance is 1,486 feet down the sheer canyon wall. It is located on government land and no visitor will be allowed there under penalty of trespass. The scientists wish to work unmolested, without fear of the archaeological discoveries being disturbed by curio or relic hunters. A trip there would be fruitless, and the visitor would be sent on his way. The story of how I found the cavern has been related, but in a paragraph: I was journeying down the Colorado River in a boat, alone, looking for mineral. Some forty-two miles up the river from the El Tovar Crystal canyon, I saw on the east wall, stains in the sedimentary formation about 2,000 feet above the river bed. There was no trail to this point, but I finally reached it with great difficulty. Above a shelf which hid it from view from the river, was the mouth of the cave. There are steps leading from this entrance some thirty yards to what was, at the time the cavern was inhabited, the level of the river. When I saw the chisel marks on the wall inside the entrance, I became interested, secured my gun and went in. During that trip I went back several hundred feet along the main passage, till I came to the crypt in which I discovered the mummies. One of these I stood up and photographed by flashlight. I gathered a number of relics, which I carried down the Colorado to Yuma, from whence I shipped them to Washington with details of the discovery. Following this, the explorations were undertaken.

The main passageway is about 12 feet wide, narrowing to nine feet toward the farther end. About 57 feet from the entrance, the first side-passages branch off to the right and left, along which, on both sides, are a number of rooms about the size of ordinary living rooms of today, though some are 30 by 40 feet square. These are entered by oval-shaped doors and are ventilated by round air spaces through the walls into the passages. The walls are about three feet six inches in thickness. The passages are chiseled or hewn as straight as could be laid out by an engineer. The ceilings of many of the rooms converge to a center. The side-passages near the entrance run at a sharp angle from the main hall, but toward the rear they gradually reach a right angle in direction.

Over a hundred feet from the entrance is the cross-hall, several hundred feet long, in which is found the idol, or image, of the people's god, sitting cross-legged, with a lotus flower or lily in each hand. The cast of the face is oriental, and the carving shows a skillful hand, and the entire is remarkably well preserved, as is everything in this cavern. The idol most resembles Buddha, though the scientists are not certain as to what religious worship it represents. Taking into consideration everything found thus far, it is possible that this worship most resembles the ancient people of Thibet [sic]. Surrounding this idol are similar images, some very beautiful in form; others crooked-necked and distorted shapes, symbolical, probably, of good and evil. There are two large cactus with protruding arms, one on each side of the dais on which the god squats. All this is carved out of hard rock resembling marble. In the opposite corner of this cross-hall were found tools of all descriptions, made of copper. These people undoubtedly knew the lost art of hardening this metal, which has been sought by chemists for centuries without result. On a bench running around the workroom was some charcoal and other material probably used in the process. There is also slag and stuff similar to matte, showing that these ancients smelted ores, but so far no trace of where or how this was done has been discovered, nor the origin of the ore.

Among the other finds are vases or urns and cups of copper and gold, made very artistic in design. The pottery work includes enameled ware and glazed vessels. Another passageway leads to granaries such as are found in the oriental temples. They contain seeds of various kinds. One very large storehouse has not yet been entered, as it is twelve feet high and can be reached only from above. Two copper hooks extend on the edge, which indicates that some sort of ladder was attached. These granaries are rounded, as the materials of which they are constructed, I think, is very hard cement. A gray metal is also found in this cavern, which puzzles the scientists, for its identity has not been established. It resembles platinum. Strewn promiscuously over the floor everywhere are what people call 'cats eyes,' a yellow stone of no great value. Each one is engraved with the head of the Malay type.

On all the urns, or walls over doorways, and tablets of stone which were found by the image are the mysterious hieroglyphics, the key to which the Smithsonian Institute hopes yet to discover. The engraving on the tablets probably has something to do with the religion of the people. Similar hieroglyphics have been found in southern Arizona. Among the pictorial writings, only two animals are found. One is of prehistoric type.

The tomb or crypt in which the mummies were found is one of the largest of the chambers, the walls slanting back at an angle of about 35 degrees. On these are tiers of mummies, each one occupying a separate hewn shelf. At the head of each is a small bench, one of which is found copper cups and pieces of broken swords. Some of the mummies are covered with clay, and all are wrapped in a dark fabric. The urns or cups on the lower tiers are crude, while as the higher shelves are reached the urns are finer in design, showing a later stage of civilization. It is worthy of note that all the mummies examined so far have proved to be male, no children or females being buried here. This leads to the belief that this exterior section was the warriors' barracks.

Among the discoveries no bones of animals have been found, no skins, no clothing, no bedding. Many of the rooms are bare but for water vessels. One room, about 40 by 700 feet, was probably the main dining hall, for cooking utensils are found here. What these people lived on is a problem though it is presumed that they came south in the winter and farmed in the valleys, going back north in the summer. Upwards of 50,000 people could have lived in the caverns comfortably. One theory is that the present Indian tribes found in Arizona are descendants of the serfs or slaves of the people which inhabited the cave. Undoubtedly a good many thousands of years before the Christian era a people lived here which reached a high stage of civilization. The chronology of human history is full of gaps. Professor Jordan is much enthused over the discoveries and believes that the find will prove of incalculable value in archaeological work.

One thing I have not spoken of, may be of interest. There is one chamber the passageway to which is not ventilated, and when we approached it a deadly, snaky smell struck us. Our lights would not penetrate the gloom, and until stronger ones are available we will not know what the chamber contains. Some say snakes, but others boo-hoo this idea and think it may contain a deadly gas or chemicals used by the ancients. No sounds are heard, but it smells snaky just the same. The whole underground installation gives one of shaky nerves the creeps. The gloom is like a weight on one's shoulders, and our flashlights and candles only make the darkness blacker. Imagination can revel in conjectures and ungodly day-dreams back through the ages that have elapsed till the mind reels dizzily in space."

In connection with this story, it is notable that among the Hopi Indians the tradition is told that their ancestors once lived in an underworld in the Grand Canyon till dissension arose between the good and the bad, the people of one heart and the people of two hearts. Machetto, who was their chief, counseled them to leave the underworld, but there was no way out. The chief then caused a tree to grow up and pierce the roof of the underworld, and then the people of one heart climbed out. They tarried by Paisisvai (Red River), which is the Colorado, and grew grain and corn. They sent out a message to the Temple of the Sun, asking the blessing of peace, good will and rain for the people of one heart. That messenger never returned, but today at the Hopi villages at sundown can be seen the old men of the tribe out on the housetops gazing toward the sun, looking for the messenger. When he returns, their lands and ancient dwelling place will be restored to them. That is the tradition. Among the engravings of animals in the cave is seen the image of a heart over the spot where it is located. The legend was learned by W.E. Rollins, the artist, during a

> year spent with the Hopi Indians. There are two theories of the origin of the Egyptians. One is that they came from Asia; another that the racial cradle was in the upper Nile region. Heeren, an Egyptologist, believed in the Indian origin of the Egyptians. The discoveries in the Grand Canyon may throw further light on human evolution and prehistoric ages.[29]

There are a number of very important points to note about this article:

1) Note the sheer amount of detail provided about the alleged discovery;
2) Note the alleged involvement of the Smithsonian Institution;
3) Note the alleged involvement of the cavern's discoverer, G.E. Kinkaid, with the Smithsonian; and finally,
4) Note the alleged chief of the Smithsonian archaeological team, Professor S.A. Jordan.

These details have become the heart of what has been something of a minor controversy over the story ever since its first appearance. Was the story true? Or was it an elaborate hoax?

David Hatcher Childress certainly belongs to the class of those who think it's true, but with reservations. As Childress quips at the beginning of his article, "It was like the plot out of a fantasy Western movie..."[30] But there were questions:

> What became of the artifacts? What became of Jordan? Did he return to the Smithsonian in Washington D.C. and disappear with all the records of his discovery? Has there been some archeological cover-up reminiscent of the last scene in the movie *Raiders of the Lost Ark* where the Ark of the Covenant is placed inside a crate in a giant warehouse never to be seen again?
>
> It has also been suggested that while the discovery was real, the archeologists working for the Smithsonian were not. These men may not have been working for the Smithsonian Institute [sic] out of Washington D.C. at all, but merely claiming to be doing so. This may have been a cover-up for an illegal archeological dig that was raiding the ancient site and claiming legitimacy from a very distant institution. It could have been very difficult indeed, in 1909, to check

29 Cited in David Hatcher Childress, "The Egyptian City of the Grand Canyon," *World Explorer Magazine,* Vol. 4, No. 8, pp. 22–28, emphasis added.

30 David Hatcher Childress, "The Egyptian City of the Grand Canyon," *World Explorer Magazine,* Vol. 4, No. 8, p. 22.

on the credentials of the archeologists. These men may well have disappeared shortly after the article appeared, but not to Washington D.C. as we might suppose, but rather to San Francisco, Los Angeles, or Denver.[31]

Oldest Paper in Phoenix—Twenty-Ninth Year.

EXPLORATIONS IN GRAND CANYON

Mysteries of Immense Rich Cavern Being Brought to Light.

JORDAN IS ENTHUSED

Remarkable Finds Indicate Ancient People Migrated From Orient.

The latest news of the progress of the explorations of what is now regarded by scientists as not only the oldest archaeological discovery in the United States, but one of the most valuable in the world, which was mentioned some time ago in the Gazette, was brought to the city yesterday by G. E. Kinkaid, the explorer who found the great underground citadel of the Grand Canyon during a trip from Green river, Wyoming, down the Colorado, in a wooden boat, to Yuma, several months ago. According to the story related yesterday to the Gazette by Mr. Kinkaid, the archaeologists of the Smithsonian Institute, which is financing the explorations, have made discoveries which almost conclusively prove that the race which inhabited this mysterious cavern, hewn in solid rock by human hands, was of oriental origin, possibly from Egypt, tracing back to Ramses. If their theories are borne out by the translation of the tablets engraved with hieroglyphics, the mystery of the prehistoric peoples of North America, their ancient arts, who they were and whence they came, will be solved. Egypt and the Nile, and Arizona and the Colorado will be linked by a historical chain running back to ages which staggers the wildest fancy of the fictionist.

A Thorough Investigation.

Under the direction of Prof. S. A. Jordan, the Smithsonian institute is now prosecuting the most thorough explorations, which will be continued until the last link in the chain is forged. Nearly a mile underground, about 1480 feet below the surface, the long main feet ventilation of the cavern, the steady draught that blows through, indicates that it has another outlet to the surface.

Mr. Kinkaid's Report.

Mr. Kinkaid was the first white child born in Idaho and has been an explorer and hunter all his life, thirty years having been in the service of the Smithsonian institute. Even briefly recounted, his history sounds fabulous, almost grotesque.

"First, I would impress that the cavern is nearly inaccessible. The entrance is 1486 feet down the sheer canyon wall. It is located on government land and no visitor will be allowed there under penalty of trespass. The scientists wish to work unmolested, without fear of the archaeological discoveries being disturbed by curio or relic hunters. A trip there would be fruitless, and the visitor would be sent on his way. The story of how I found the cavern has been related, but in a paragraph: I was journeying down the Colorado river in a boat, alone, looking for mineral. Some forty-two miles up the river from the El Tovar Crystal canyon I saw on the east wall, stains in the sedimentary formation about 2000 feet above the river bed. There was no trail to this point, but I finally reached it with great difficulty. Above a shelf which hid it from view from the river, was the mouth of the cave. There are steps leading from this entrance some thirty yards to what was, at the time the cavern was inhabited, the level of the river. When I saw the chisel marks on the wall inside the entrance, I became interested, secured my gun and went in. During that trip I went back several hundred feet along the main passage, till I came to the crypt in which I discovered the mummies. One of these I stood up and photographed by flashlight. I gathered a number of relics, which I carried down the Colorado to Yuma, from whence I shipped them to Washington with details of the discovery. Following this, the explorations were undertaken.

The Passages.

"The main passageway is about 12 feet wide, narrowing to 9 feet toward the farther end. About 57 feet from the entrance, the first side-passages branch off to the right and left, along which, on both sides, are a number of rooms about the size of ordinary living rooms of today, though some are 30 or 40 feet square. These are entered by oval-shaped doors and are ventilated by round air spaces through the walls into the passages. The walls are about 3 feet 6 inches in thickness. The passages are chiseled or hewn as straight as could be laid out by an engineer. The ceilings of many of the rooms converge to a center. The side-passages near the entrance run at a sharp angle from the main hall, but toward the rear they gradually reach a right angle in direction.

The Shrine.

"Over a hundred feet from the entrance is the cross-hall, several hundred feet long, in which was found the idol, or image, of the people's god, sitting cross-legged, with a lotus flower or lily in each hand. The cast of the

April 5, 1909 *Phoenix Gazette* Article

31 Ibid., p. 28.

While Childress offers no source for his speculations that the so-called archaeologists might not have been from the Smithsonian at all, nor any source for the speculation that the "archaeologists" went to San Francisco, Los Angeles, or Denver, the real questions Childress raises nonetheless remain: were the discoveries in fact *real?* If so, what *happened* to them? In order to answer those questions, one must determine whether or not the article was genuine, or a hoax.

3. Hoax or Cover-up?

One of those questioning the entire story is Philip Coppens, who notes that the story says little about Egypt, but that it points even further east:

> [The] account is quite factual. Idols "resemble" Buddha, rather than "are" Buddha. The worship "resembles" that of Tibet, not "is"... Kinkaid is trying to use analogies to explain his discovery. It is the anonymous author of the article who makes the connection with ancient Egypt and lets his mind float to one of the biggest discoveries of all time.[32]

Coppens is suggesting, in other words, that the anonymous author of the newspaper article deliberately exaggerated the story, casting a pall of suspicion over just exactly what, if anything at all, was actually discovered. The fact that the newspaper never followed up on the sensational story raises further suspicions.[33]

But there was a further problem: the article made three explicit claims:

1) that the discoverer of the cave, G.E. Kinkaid, worked for the Smithsonian Institution;
2) that the Smithsonian Institution was thus involved in the excavation and recovery of the artifacts; and,
3) that a professor of archaeology, one S.A. Jordan, was in charge of the project.

However, as Coppens observes, when contacted in 2000 for confirmation of these allegations, the Smithsonian's reply was unequivocal:

> The Smithsonian Institution has received many questions about an article in the April 5, 1909 *Phoenix Gazette* about G.E. Kinkaid and

32 Philip Coppens, "Canyonitis: Seeing Evidence of Ancient Egypt in the Grand Canyon," www.philipcoppens.com/egyptiancanyon.html, p. 1.

33 Ibid.

> his discovery of a "great underground citadel" in the Grand Canyon, hewn by an ancient race of oriental origin, possibly from Egypt... The Smithsonian's Department of Anthropology has searched its files without finding any mention of a Professor Jordan, Kinkaid, or a lost Egyptian civilization in Arizona. Nevertheless, the story continues to be repeated in books and articles.[34]

This would seem to be the end of the story...

...but no. Coppens admits that "there is room for a cover-up, of course,"[35] simply because it is a standard practice of such organizations to deliberately "look" for certain files in the wrong place. Not finding them in a certain place does not mean that they do not exist; they merely do not exist where the search was conducted:

> The files do not necessarily have to be set within that department's and the reference to the *Phoenix Gazette* rather than *Arizona Gazette* could be a simple error, or an escape valve that is so often present in official replies engineered to debunk. Stories like "the CIA Division X has no record" often means that Division Y is the one who has that record.[36]

It is a ploy well known to researchers using FOIA requests to pry information loose from various government departments.

So the important question remains: if the entire newspaper story was a hoax, who perpetrated it? And why?

Coppens notes that if the original story *was* a hoax perpetrated by the newspaper itself — in the tradition of the yellow journalism of the day — in order to generate more sales and circulation, then it is nothing less than bizarre that it never followed up the story with a sequel. Thus it is unlikely that if the story was a hoax, that the *Arizona Gazette* was the perpetrator, which leaves Kinkaid himself.[37] A number of minor discrepancies exist in Kinkaid's account that, while not proving that he perpetrated a hoax, at least raise questions.[38] "So where does that leave us?" Coppens asks. His answer: somewhere in the middle, between a hoax and an exaggerated truth. In Coppen's opinion,

34 Philip Coppens, "Canyonitis: Seeing Evidence of Ancient Egypt in the Grand Canyon," www.philipcoppens.com/egyptiancanyon.html, p. 1.

35 Ibid.

36 Ibid.

37 Philip Coppens, "Canyonitis: Seeing Evidence of Ancient Egypt in the Grand Canyon," www.philipcoppens.com/egyptiancanyon.html, p. 2.

38 Ibid.

the discovery probably had more to do with the discovery of artifacts of the ancient Anasazi American Indian culture.[39]

This solution, however, raises as many questions as it answers. For one thing, the Anasazi culture was known to anthropology and archaeology at the time. So why cover it up? The Smithsonian's silence and claims of non-involvement are equally suspicious, for it would have a natural and inherent interest in any *reporting* of such a story, and would inevitably have investigated. This makes its blatant denials of any involvement highly suspicious.

In an article entitled "Archaeological Coverup?" Jason Colavito makes the comment — in support of the newspaper story being a complete hoax — that the article refers to the Smithsonian *Institute* and not the Smithsonian *Institution,* and that anyone claiming to work for it would know this. Colavito also argues that the original story in the *Arizona Gazette* is a single-source story without any other external corroboration, and thus argues that the whole thing is a likely hoax.[40]

Childress, however, points out that the amount of specific detail in the article plus the fact that it ran in an otherwise ordinary newspaper argues against the story being a hoax. But for Childress the story is part of a much *larger* picture, one of an "archaeological cover-up" of anomalous evidence and artifacts.

Childress notes that

> To those who investigate allegations of archaeological cover-ups there are disturbing indications that the most important archaeological institute in the United States, the Smithsonian Institute [sic], an independent federal agency, has been actively suppressing some of the most interesting and important archaeological discoveries made in the Americas.[41]

There is an obvious fact here that is often overlooked: the U.S. government, like the Nazi Reich long afterward, and like many other governments of the major powers, established an official agency to do nothing but sponsor archaeological research. Thus, it stands to reason that like any other government agency, the Smithsonian would have its own "classified secrets." The question is, why would such a seemingly staid and sober discipline and institution be keeping secrets?

---

39 Ibid., p. 3.

40 Q.v. Jason Colavito, "Archaeological Coverup?" jcolavito.tripod.com/lostcivilizations/id8.html, p. 1.

41 David Hatcher Childress, "Archeological Coverups" www.unexplainable.net/artman/publish/article_1727.shtml, p. 2.

For Childress, the answer is rather simple: the Smithsonian had an agenda to promote a particular historical and anthropological paradigm, and to suppress evidence of another:

> The cover-up and alleged suppression of archaeological evidence began in late 1881 when John Wesley Powell, the geologist famous for exploring the Grand Canyon, appointed Cyrus Thomas as the director of the Eastern Mound Division of the Smithsonian Institution's Bureau of Ethnology.
>
> When Thomas came to the Bureau of Ethnology he was a "pronounced believer in the existence of a race of Mound Builders, distinct from the American Indians." However, John Wesley Powell, the director of the Bureau of Ethnology, a very sympathetic man toward the American Indians, had lived with the peaceful Winnebago Indians of Wisconsin for many years as a youth and felt that American Indians were unfairly thought of as primitive and savage.
>
> The Smithsonian began to promote the idea that the Native Americans... were descended from advanced civilizations and were worthy of respect and protection.
>
> They also began a program of suppressing any archaeological evidence that lent credence to the school of thought known as Diffusionism, a school which believes that throughout history there has been widespread dispersion of culture and civilization via contact by ship and major trade routes.
>
> The Smithsonian opted for the opposite school, known as Isolationism. Isolationism holds that most civilizations are isolated from each other and that there has been very little contact between them, especially those that are separated by bodies of water. In this intellectual war that started in the 1880s, it was held that even contact between the civilizations of the Ohio and Mississippi Valleys were rare, and certainly these civilizations and did not have any contact with such advanced cultures as the Mayas, Toltecs, or Aztecs in Mexico and Central America.[42]

But according to Childress, in addition to suppressing evidence supporting the Diffusionism hypothesis, the Smithsonian suppressed evidence of a very different sort...

42 David Hatcher Childress, "Archeological Coverups" www.unexplainable.net/artman/publish/article_1727.shtml, p. 2.

a. Destroyed and Suppressed Evidence: Curious NASA Parallels

That evidence concerned the existence of the giants mentioned in ancient texts:

> When the contents of many ancient mounds and pyramids of the Midwest were examined, it was shown that the history of the Mississippi River Valleys was that of an ancient and sophisticated culture that had been in contact with Europe and other areas. Not only that, *the contents of many mounds revealed burials of huge men, sometimes seven or eight feet tall,* in full armour with swords and sometimes huge treasures.
>
> ...
>
> For instance, when Spiro Mound in Oklahoma was excavated in the 1930s, a tall man in full armour was discovered along with a pot of thousands of pearls and other artefacts [sic], the largest such treasure so far documented. The whereabouts of the man in armour is unknown and it is quite likely that it eventually was taken to the Smithsonian Institution.[43]

The discovery of large "human" remains tends to support ancient Native American Indian legends that asserted the existence of "monsters and giants" with whom the Indians had once lived, legends which we will encounter later.

At this juncture, Childress recounts a story that an anonymous researcher related to him, namely that the Smithsonian had dismissed one of its senior employees for defending the "heresy" of Diffusionism, and for attempting to blow the whistle on an operation where the Institution loaded a barge with anomalous evidence and literally sank it in the Atlantic to hide the evidence.[44] While Childress offers no supporting evidence for this story, it is curious that it *does* fit a wider pattern of the suppression of anomalous evidence of ancient sophisticated civilizations and their artifacts uncovered by another government agency: NASA. As space anomalies researchers Richard C. Hoagland and Mike Bara have pointed out, NASA's own version of the "Smithsonian barge" included such evidence-suppressing tactics as blacked-out thumbnail pictures in the NASA Apollo photograph catalogues,[45] airbrushing and oth-

---

43 David Hatcher Childress, "Archeological Coverups" www.unexplainable.net/artman/publish/article_1727.shtml, p. 3, emphasis added.

44 Ibid.

45 Richard C. Hoagland and Mike Bara, *Dark Mission: The Secret History of NASA* (Port Townsend, Washington: Feral House, 2007), p. 121.

erwise doctoring Apollo photos,[46] and — incomprehensibly — orders to Apollo archivists to actually *destroy* priceless Apollo photo archives, an order which was fortunately disregarded by one individual.[47] All of this evidence, as Hoagland and Bara observe, was strongly suggestive of the presence of a high and technologically sophisticated civilization on the Moon and Mars. NASA was, in effect, following the same paradigm as Childress suggested the Smithsonian was following: suppressing evidence of Diffusionism, only in this case, it was the ultimate in Diffusionism: the dispersal of civilization not only on planet Earth, but her neighbors as well.

If there is a common motivation here between the two government agencies, then I am bold to suggest that it lies in the common factor they each share, namely that the existence of technological artifacts, of the skeletal remains of giants, and other anomalies, tended to corroborate that which ancient texts and traditions had always maintained: there once was such a sophisticated civilization, that mankind himself is its deliberately engineered creation, and there once were such things as giants. In short, anomalous evidence from space, and archaeological evidence from the Earth, that tended to corroborate such ancient myths, had to be suppressed. The question is, *why?* The answer will be encountered in the final chapter.

b. The Smithsonian and the Suppression of the Alaskan Giants

There is another episode recounted by Childress that loosely corroborates the allegations of the Smithsonian's suppression of anomalous archaeological evidence:

> Ivan T. Sanderson, a well-known zoologist and frequent guest on Johnny Carson's *Tonight Show* in the 1960s... once related a curious story about a letter he received regarding an engineer who was stationed on the Aleutian island of Shemya during World War II. While building an airstrip, his crew bulldozed a group of hills and discovered under several sedimentary layers what appeared to be human remains. The Alaskan mound was in fact a graveyard of gigantic human remains, consisting of crania and long leg bones.
>
> The crania measure from 22 to 24 inches from base to crown. Since an adult skull normally measures about eight inches from back to front, such a large cranium would imply an immense size for a

46 Ibid., pp. 146–147.

47 Richard C. Hoagland and Mike Bara, *Dark Mission: The Secret History of NASA* (Port Townsend, Washington: Feral House, 2007), pp. 144–153.

normally proportioned human. Furthermore, every skull was said to have been neatly trepanned (a process of cutting a hole in the upper portion of the skull).

In fact, the habit of flattening the skull of an infant and forcing it to grow in an elongated shape was a practice used by ancient Peruvians, the Mayas, and the Flathead Indians of Montana. Sanderson tried to gather further proof, eventually receiving a letter from another member of the unit who confirmed the report. The letters both indicated that the Smithsonian Institution had collected the remains, yet nothing else was heard. Sanderson seemed convinced that the Smithsonian Institution had received the bizarre relics, but wondered why they would not release the data. He asks, "...is it that these people cannot face rewriting all the textbooks?"[48]

Notwithstanding the weighty credentials of Sanderson, himself a well-known investigator of Fortean phenomena, many, however, still doubt the existence of such coordinated archaeological cover-ups.

4. Archaeology-gate: Cremo, Thompson, and the Antiquity of Man

Unfortunately for such individuals, decidedly more weighty evidence of such archaeological cover-ups was presented by Michael A. Cremo and Richard L. Thompson in their massive and magisterial study of the antiquity of mankind and his artifacts, *Forbidden Archeology: The Hidden History of the Human Race.* Any attempt to summarize this massive and scholarly tome of over nine hundred pages here is simply impossible. We may, however, get some idea of their thesis, method, and the vast amount of evidence that they present to prove that anatomically modern tool-making humans have been around for far longer than standard scientific dogma proclaims.

They begin by observing that "...'[Facts]' turn out to be networks of arguments and observational claims."[49] Thus, "a piece of evidence is anomalous only in relation to a particular theory,"[50] the theory in this case being evolution and all its implications for anthropology and the presumed reconstruction of the history and evolution of mankind. In this context that they ask the question that introduces the nine hundred pages of "anomalous" evidence that does not fit the standard theory:

---

48 David Hatcher Childress, "Archeological Coverups" www.unexplainable.net/artman/publish/article_1727.shtml, p. 4.

49 Michael A. Cremo and Richard L. Thompson, *Forbidden Archeology: The Hidden History of the Human Race* (Los Angeles: Bhaktivedanta Book Publishing Inc., 1996), p. 19.

50 Ibid., p. 23.

> What if, for example, fossils of anatomically modern humans turned up in strata older than those in which the dryopithecine apes were found? Even if anatomically modern humans were found to have lived contemporaneously with *Dryopithecus* (or even a million years ago, 4 million years after the Late Miocene disappearance of *Dryopithecus*), that would be enough to throw the current accounts of the origin of humankind completely out the window.
>
> In fact, such evidence has already been found, **but it has since been suppressed or conveniently forgotten...**
>
> ... Before Java man... reputable nineteenth-century scientists found a number of examples of anatomically modern human skeletal remains in very ancient strata.[51]

They then state that there is "an essential equivalence" in the evidence adduced in favor of the standard theory, and the anomalous evidence against it.

This has repercussions, for "it is not appropriate to accept one and reject the other." Indeed,

> If we reject the first set of reports (the anomalies) and to be consistent, also reject the second set (evidence currently accepted), then the theory of human evolution is deprived of a good part of its observational foundation. But if we accept the first set of reports, then we must accept the existence of intelligent, toolmaking beings in geological periods as remote as the Miocene, or even the Eocene. If we accept the skeletal evidence presented in these reports, we must go further and accept the existence of anatomically modern human beings in these remote periods.[52]

In addition to a thorough review of human skeletal remains in anomalously *old* strata, Cremo and Thompson produce a vast catalogue of extremely anomalous objects that are the result of obvious art and intelligence. These included carved marble figures in strata that "suggests the characters were made by intelligent humans from the distant past,"[53] a section of gold thread found

51 Ibid., pp. 18–19, italics in the original, boldface emphasis added.

52 Michael A. Cremo and Richard L. Thompson, *Forbidden Archeology: The Hidden History of the Human Race* (Los Angeles: Bhaktivedanta Book Publishing Inc., 1996), p. 22.

53 Michael A. Cremo and Richard L. Thompson, *The Hidden History of the Human Race: The Condensed Version of Forbidden Archeology* (Los Angeles: Bhaktivedanta Book Publishing, 1999), p. 105.

in strata between 320 and 360 million years old,[54] a report in a nineteenth-century edition of *Scientific American* recording the discovery of a metallic vase in strata 600 million years old,[55] a chalk ball in France in strata 45–55 million years old,[56] a machined coin with undecipherable writing at least 200,000 years old, discovered in Illinois,[57] a clay figurine discovered in Idaho that is at least two million years old.[58] The list of suppressed and conveniently forgotten discoveries goes on and on, and we shall have occasion to refer to this list again in the final chapter, and in a very different context.

So why suppress or "conveniently forget" such evidences?

In their remarks cited previously, Cremo and Thompson give one obvious answer: such evidence does not serve the reigning scientific dogma of evolution.

It is when Cremo and Thompson's careful scholarship is viewed within the context of deeper allegations of archaeological suppression that other more disturbing philosophical and speculative implications emerge, for the bottom line is that both the archaeological suppression of "convenient forgetting" and the dogma of evolution would seem to be tailor-made devices to suppress the ideas of the great antiquity of man and the possibility of a previously existing Very High Civilization. Viewed in this light, it is possible that the scientific theory functions as an "evidence screening mechanism" for scientists *not* privy to the ultimate agenda: the suppression of the very notion of such a civilization, and therefore of any attempt to investigate it by the hoi polloi. One might indeed be dealing with a "public consumption" biology, anthropology, and archaeology to parallel the "public consumption" physics deliberately designed to prevent the recovery of lost technologies and science. Clearly, if our examination of the I Ching and DNA are any indicator, this would appear to be the case.

5. An Aside: Gilgamesh Discovered — Speculative Implications

In this context, an interesting story was reported by the BBC on Tuesday, April 29, 2003. German archaeologists in Iraq located the site of ancient Uruk, including, they believed, the tomb of its famous king, Gilgamesh.[59] After this initial reporting, little was ever mentioned about the find. Of course, this could be due to modern journalism's laziness in following up a story.

54 Ibid., p. 106.

55 Ibid., pp. 106–107.

56 Ibid., pp. 107–109.

57 Ibid., pp. 109–110.

58 Ibid., pp. 110–113.

59 "Gilgamesh Tomb Believed Found," BBC News, Tuesday, 29 April 2003, 8:578 GMT, news.bbc.co.uk/2hi/science/nature/2982891.stm, p. 1.

But there could be something more, for it will be recalled that the ancient texts state that Gilgamesh was "two thirds god and one third human." That is, Gilgamesh was himself a chimerical hybrid offspring of the gods and of man, and if one is to believe the suggestions of ancient genetic engineering explored previously, it could be possible that via modern techniques of sequencing DNA from cadavers, scientists might actually be looking for that "divine component," and Gilgamesh's remains would offer the perfect means to do so.

But don't hold your breath. If they are doing that, chances are they will not tell us what they find.

D. Conclusions: Taking Stock Thus Far

So, now, what do we have? What sort of speculative conclusions might we draw from this brief essay of genome wars and connections of miscellanies? Of ancient measures and hidden elites?

1) We have, first, the presence of elites manipulating information in both modern and ancient times, and while it does not follow from the evidence discussed that these elites are connected or continuous throughout time, it is clear that the methods and motivations are oftentimes eerily parallel;
2) In the case of the elites behind the establishment of a system of geodetic- and astronomically-based measures over a broad area from Britain to Mesopotamia, it is clear that one possible agenda behind this activity was to jump-start civilization back into existence, by fostering international trade. In this respect, as I demonstrated in my previous book *Babylon's Banksters*, there is a clear association of ancient temple priesthoods both with astronomical measuring activity, with astrological predictive or divination activity, and with the issuance of money, activities that are in turn deeply connected to each other in the deep relationship between physical and economic systems;
3) In the case of the ancient texts examined, it is clear that mankind is an engineered being, created for the explicit purpose of servitude by a genetically related species, which may accordingly have a "legal" claim upon humanity;
4) It is clear that the modern elite, for various reasons, are attempting to suppress this aspect of human development and origins, in part in service to evolutionary theory, but possibly because there is a deeper agenda, which would include obscuring possible

human origins and the implied existence of technological sophistication in paleoancient times, and their attempts both to recover it, and to reconstruct a true but hidden account of human history;

5) It is clear from the comparison of modern genetic discoveries with the ancient Chinese system of divination known as the I Ching that someone had refined, precise scientific data concerning genetic structure, and that somehow this was in turn connected or conceived to be an intimate clue into the nature of the physical processes of the universe itself. In a certain sense, then, the I Ching might be called the "other half" of DNA, a fragment of a once-lost but highly unified scientific worldview whose views about God, Man, and the Medium were highly unified, formally explicit, and required no acts of faith to comprehend;
6) The existence of genetic science also affords the modern elites the method and basis to verify some of the claims of the ancient texts, though such verification is likely to be performed covertly and its results kept secret and known only to that elite.

With this in mind, it is time, finally, to turn to the two remaining components that form the themes of this essay: monsters, and men.

III.

Monsters and Men

"The Natives' story was consistent: these were vestiges of a giant race, now extinct owing to natural catastrophe or battles with humans of the distant past."

—Adrienne Mayor,

Fossil Legends of the First Americans, p. 78.

"What the mitochondrial gene tree did was to introduce an objective time-depth measurement into the equation for the first time. It showed quite clearly that the common mitochondrial ancestor of all modern humans lived only about 150,000 years ago."

—Bryan Sykes,

The Seven Daughters of Eve: The Science that Reveals Our Genetic Ancestry, p. 50.

"Even if anatomically modern humans were found to have lived... even a million years ago... that would be enough to throw the current accounts of the origin of humankind completely out the window. In fact, such evidence has already been found, but it has since been suppressed or conveniently forgotten."

—Michael A. Cremo and Richard L. Thompson,

Forbidden Archeology: The Hidden History of the Human Race, p. 18.

Seven

The Return of the Sirrush:

Greeks, Indians, Giants and Monsters

∴

"Keep in mind that in many Native traditions, 'giants' of ancient eras were often understood to be primeval beings that were neither animal nor human."
—Adrienne Mayor[1]

GENETIC ENGINEERING HAS BECOME something of a fascination within the community of researchers studying alternative, "Fortean" phenomena and perspectives on history and ethics. Stories abound of the genetic alteration of food and seeds, with agribusiness companies like Monsanto leading the charge for the ridiculous idea of seeds for plants that will not reproduce... a nice way to control the food supply. We are reminded of the great potentials for good in genetic cures for the ravages of cancer, AIDS, diabetes and a host of other ailments, and of the great potential for evil in the genetic engineering of bio-weapons that target certain population groups with specific biological markers. Other stories recount the creation of what can only be described as science fiction: rabbits and rats and cats that glow in the dark, mice growing human ears, human DNA put into ordinary food products or artificial sweeteners or spliced into the genes of cattle and pigs.

If we did not know better, we would think we were reading the ancient Mesopotamian "creation" epic, the *Enuma Elish,* with its account of "scorpion men," or other ancient tales of Mesopotamia of the "fish god" Oannes,

1 Adrienne Mayor, *Fossil Legends of the First Americans* (Princeton University Press, 2005), p. 36.

part fish and part man, and of other chimerical hybrids recounted in ancient texts and of the chimerical gods depicted in Egyptian and Mesopotamian art. When Alexander the Great's empire was divided among his generals — with Mesopotamia going to the Seleucids and Egypt to the Ptolemies — each of those new dynasties commissioned "official histories" of their new realms in Greek, and solicited experts from within their realms to translate from their archives the essential texts. One of these was the Babylonian priest, scribe, and historian Berossus, whose *Babylonaica* survives only in fragmented quotations from his work cited by later classical historians and church fathers.

According to the early church historian Eusebius, who in turn is citing Alexander Polyhistor, Berossus stated:

> There was a time in which there existed nothing but darkness and an abyss of waters, wherein resided *most hideous beings, which were produced of a two-fold principle. There appeared men, some of whom were furnished with two wings, others with four, and with two faces. They had one body but two heads: one that of a man, the other of a woman; and likewise in their several organs both male and female. Other human figures were to be seen with the legs and horns of goats: some had horses' feet; while others united the hind quarters of a horse with the body of a man, resembling in shape the hippocentaurs. Bulls likewise were bred there with the heads of men; and dogs with fourfold bodies, terminated in their extremities with the tails of fishes; horses also with the heads of dogs; men too and other animals with the heads and bodies of horses and the tails of fishes.* In short, there were creatures in which were combined the limbs of every species of animals. In addition to these, fishes, reptiles, serpents, with other monstrous animals, which assumed each other's shape and countenance. Of all which were preserved delineations in the temple of Belus at Babylon.
>
> The person who presided over them was a woman named Omoroca; which in the Chaldaean language is Thalatth; in Greek Thalassa, the sea; but which might equally be interpreted the Moon. All things being in this situation, Belus came, and cut the woman asunder; and of one half of her he formed the earth, and of the other half the heavens; and at the same time destroyed the animals within her. All this (he says) was an allegorical description of nature. For, the whole universe consisting of moisture, and animals being continually generated therein, *the deity above-mentioned took off his own head; upon which the other gods mixed the blood, as it gushed out, with the earth; and from thence were formed men.* On this account it

> is that they are rational, and partake of divine knowledge.... (Such, according to Polyhistor Alexander, is the account which Berossus gives in his first book.)[2]

Before analyzing this important passage, there is another fragment of Berossus that must be mentioned, also a fragment from Alexander Polyhistor that survives in the early church historian Eusebius:

> (In the second book was contained the history of the ten kings of the Chaldaeans, and the period of the continuance of each reign, which consisted collectively of a hundred and twenty sari, or four hundred and thirty-two thousand years; reaching to the time of the Deluge. For Alexander, enumerating the kings from the writings of the Chaldaeans, after the ninth Ardates, proceeds to the tenth, who is called by them Xisuthrus, in this manner:)
>
> After the death of Ardates, his son Hizuthrus reigned eighteen sari. In his time happened a great Deluge; the history of which is thus described. *The Deity, Cronus, appeared to him in a vision, and warned him that upon the fifteenth day of the month Daesius there would be a flood, by which mankind would be destroyed. He therefore enjoined him to write a history of the beginning, procedure, and conclusion of all things; and to bury it in the city of the Sun at Sipara;* and to build a vessel, and take with him into it his friends and relations; and to convey on board every thing necessary to sustain life, together with all the different animals both birds and quadrupeds, and trust himself fearlessly to the deep. *Having asked the Deity, whither he was to sail? he was answered, "To the Gods";* upon which he offered up a prayer for the good of mankind. He then obeyed the divine admonition: and built a vessel five stadia in length, and two in breadth. Into this he put every thing which he had prepared; and last of all conveyed into it his wife, his children, and his friends.
>
> After the flood had been upon the earth, and was in time abated, Xisuthrus sent out birds from the vessel; which, not finding any food, nor any place whereupon they might rest their feet, returned to him again. After an interval of some days, he sent them forth a second time; and they now returned with their feet tinged with mud. He made a trial a third time with these birds; but they returned to him no more; from whence he judged that the surface of the earth had appeared above the waters. He therefore made an opening in the vessel, and upon look-

2 Fragments from Berossus, www.sacred-texts.com/cla/af/af02.htm, pp. 1–2, emphasis added.

> ing out found that it was stranded upon the side of some mountain; upon which he immediately quitted it with his wife, his daughter, and the pilot. Xisuthrus then paid his adoration to the earth; and having constructed an altar, offered sacrifices to the gods, and, with those who had come out of the vessel with him, disappeared.
>
> They, who remained within, finding that their companions did not return, quitted the vessel with many lamentations, and called continually on the name of Xisuthrus. Him they saw no more; but they could distinguish his voice in the air, and could hear him admonish them to pay due regard to religion; and likewise informed them that it was upon account of his piety that he was translated to live with the gods; that his wife and daughter, and the pilot, had obtained the same honour. *To this he added, that they should return to Babylonia; and, as it was ordained, search for the writings at Sippara, which they were to make known to all mankind....*[3]

The sirrush, recounted in chapter one, not only comes to mind reading these texts, but with the advent of genetic engineering techniques, has a very real possibility of coming to life.

And that's the point, for if certain ancient texts can be interpreted along technological lines and give hints and clues as to the existence of just precisely such a genetic engineering technology in high antiquity, even being used in the engineered creation of mankind himself as was seen in a previous chapter, then the horrible possibility arises that in reading about those ancient accounts of chimeras and monsters, we might not be reading ancient science fiction at all, but a reality, dimly remembered and passed down through the ages.

However, there are a number of points to notice about these fragments from Berossus that are quite important, for they play directly not only to the interpretive method one brings to ancient texts, but, if one takes them seriously, raise many chronological problems that will be dealt with not only here, but much more extensively in the next and final chapter. These points may be summarized as follows:

1) As noted, Berossus refers to creatures "of a two-fold principle," i.e., to chimerical hybrids. On the face of it, these are either the offspring of the mytheological imagination, *or,* if one grants the existence of a sophisticated technological society in high antiquity, were possibly the products of genetic engineering;

3 Fragments from Berossus, www.sacred-texts.com/cla/af/af02.htm, pp. 3–4, emphasis added.

2) These creatures were in most instances a mixture of species, though in one special instance, were masculo-androgynous mixtures of male and female humans, an odd statement for an ancient text to make, since males carry both sexual chromosomes;
3) They were associated with "the sea," a primordial abyss of waters that the Greeks called "Thalassa." In mythological terms, this is none other than the goddess Tiamat;
4) Tiamat, who created these "productions," was destroyed by a god named "Belus," and from her remains the modern heaven and earth were fashioned. Thus, Berossus is referring to the ancient war between Tiamat and Marduk, recounted in the ancient Mesopotamian "war epic," the *Enuma Elish;*[4]
5) Mankind is specifically stated to be himself a chimerical creation from the blood of a "god" (who had removed his head for the purpose!) and the earth, dimly recalling the O'Briens' technological interpretation of the Kharsag tablets in chapter five, a fact that, if tested scientifically, should show up in the genetic markers of humanity (not that we'll ever be told about it!);
6) The "Noah" character of the second passage cited — Xisuthrus — takes not only his own family but his *friends* into his ark along with all the animals. His family is specifically stated to have been "translated" to the gods, leaving the remainder of the human family to descend from the friends he took with him;
7) Prior to entering the Ark, Xisuthrus is commanded to write the *antediluvian history* of mankind and deposit it at Siparra, and upon conclusion of the episode, his companions and friends return there to recover and disseminate that history;
8) The second passage cited makes specific reference to the god "Cronus," the ancient Greeks' name for the planet *Saturn*, and Cronus, when queried by Xisuthrus as to where he is to pilot his ark, is answered "to the gods," which, given the planetary reference, means *off the "earth" entirely*. This extraterrestrial, interplanetary context is confirmed in loose fashion by Xisuthrus' family being "translated to the gods" after the deluge;
9) The Deluge is stated to have occurred some four hundred and thirty-two thousand years ago, *long before the rise of the societies recording the event*, and, as we shall see from the genetic evidence

4 For the interpretation of this epic as a *war* epic and not a *creation* epic (the standard academic view), and as a story containing many technological clues both to the existence of a sophisticated physics and genetics technology in ancient times, see my *The Giza Death Star Destroyed*, pp. 37–49, and my *The Cosmic War*, pp. 1540–162.

to be presented in the next chapter, long before the rise of modern *Homo sapiens sapiens;*

10) The war of the gods (Tiamat and Marduk, or Thalassa and Belus as they are called in Berossus), occurred prior to the deluge, and moreover, a subsequent war was associated with some "Tower of Babel moment": "They say that the *first inhabitants* of the earth, glorying in their own strength *and size,* and in despising the gods, undertook to raise a tower whose top should reach the sky, and in the place in which Babylon now stands; but when it approached the heaven, the winds assisted the gods, and overthrew the work upon its contrivers; and its ruins are said to be still at Babylon: and the gods introduced a diversity of tongues among men, who till that time had all spoken the same language; *and a war arose between Cronus and Titan.*"[5] Notably, the language of this fragment of Berossus is oddly ambiguous, for the first inhabitants of the earth are referred to not as "men" but merely "inhabitants," and then he subsequently refers to them as "men." These "first inhabitants" are also referred to as glorying in their "size," a statement which lends credence to the idea that they were of large stature. Finally reference is made to the war between Cronus (Saturn), and Titan, who of course in the Greek version of this war sired a race of giants.

11) It is to be noted that some of the references in Berossus (as well as in the *Enuma Elish)* are astrological: e.g. "scorpion men" in the *Enuma Elish,* "centaurs" in Berossus.

Of course, all this is rather breathtaking stuff... *if* one reads it "literally" and as containing glimmers of a technological past only dimly remembered. If read that way it presents numerous chronological and scientific difficulties, as we shall discover in the next and final chapter.

What is important to note here is the *agenda* that these ancient texts suggest is in place in *modern* times, for with modern experiments in "manimals" and the creation of other chimerical creatures, it would appear as a possibility that *someone* has an agenda to (re-)create the chimerical creatures of ancient myth and astrological lore. If so, then the one possible motivation for doing so is also perhaps suggested by those ancient myths: someone is attempting to create the conditions and lost technology of the ancient war against the "giants and monsters."

5 Fragments from Berossus, www.sacred-texts.com/cla/af/af02.htm, p. 9, emphasis added.

But there is the other possibility of interpretation of these ancient texts, the mythological one, and its own problematical difficulties may be revealed by asking a simple question: What did the ancients *themselves* think when they encountered the fossilized bones of large humanoid beings, or of dinosaurs? To answer this question, we must travel from Mesopotamia to Greece, and from there, across the ocean to North America, where two unlikely traditions — the Greek and the Native American Indian — give oddly parallel answers, answers that *ultimately,* and if read carefully, do not support a standard academic and "mythological" interpretation. As this careful comparison will reveal, there are astonishing similarities of traditions from places in the world otherwise disconnected from each other.

A. Greeks, Giants, Monsters, and War

One scholar who has been collecting precisely such accounts, legends and traditions, and doing so with a great deal of skill and analytical thoroughness, is Adrienne Mayor. Her works, *Fossil Legends of the First Americans* and *The First Fossil Hunters: Paleontology in Greek and Roman Times*, are not only the only such thorough studies, they are magisterial catalogues of obscure references and traditions all but forgotten to paleontological science.[6]

1. The Gigantomachy, or the War Against the Giants

The best place to afford an entry into the subject of Graeco-Roman responses to fossil remains is in the myth of the gigantomachy, the war between the Greek gods and the Titans, or giants. The "Titan wars" or wars against the giants began when the first "supreme god" of the pantheon, Cronos (Saturn) and his consort Rhea gave birth to Zeus and his siblings. Zeus in turn, with the assistance of the Titans, overthrew Cronos as supreme god and thus ushered in the era of the Olympian gods of classical Greek mythology. Then began the "Titan Wars" or wars of the giants, as some of the older giants and monsters waged war "against Zeus and the new, more human, gods."[7] This point about monsters, giants, and "more human" gods is an important one, and we shall be returning to it again in this and the next chapter.

At this juncture, Zeus defeated the giants by throwing his lightning bolts, destroying the giants' legions and their leader, Typhon.[8] Interestingly

6 Mayor has also published a fascinating study of biological and chemical warfare in ancient times entitled *Greek Fire, Poison Arrows, and Scorpion Bombs: Biological and Chemical Warfare in the Ancient World* (Overlook Duckworth, 2009).

7 Adrienne Mayor, *The First Fossil Hunters,* p. 195.

8 For a fuller discussion of the possible scientific bases behind the ancient myths of the "divine

enough, the locations on earth where these titanic battles were alleged to have occurred are often found near fossil fields dotted throughout the eastern Mediterranean, and particularly in Greece.[9] In other words, the ancients explained their encounters with fossils *by interpreting them from the standpoint of their preexisting myths.*

Interestingly enough, however, the key to the giants is their chimerical nature:

> In early Greek art, giants were imagined as quadruped monsters, or as warriors, huge ogres, or primitive strongmen armed with tree trunks and boulders; some later artists added serpent legs to symbolize their earth-born nature. *It's important to keep in mind that giants were not necessarily visualized as human.*[10]

In other words, the Gigantomachy was a literal war, a struggle for survival between at least two different species, one an older race of "monsters and ogres," and the other the new "more human" Olympian gods. This will be an important clue in resolving the problems presented by modern genetics and ancient texts, as will be seen in the next chapter. For now, we note once again that the struggle, read at face value, is between two species, one less, and one more, "human."

Mayor cites an impressive list of classical sources attesting to the ancients' perceptions of these giants as "deformed" creatures, from Flavius Josephus,[11] Manilius,[12] to the church fathers Clement of Rome,[13] and Augustine of Hippo.[14] Two of the classical sources Mayor cites are worth mentioning in more detail, however.

Diodorus Siculus observes in his *Library* that the giants "Started a war against the gods... and were completely exterminated." The one doing the exterminating was, according to Diodorus, Heracles, a.k.a. Aries, Mars, Errakal, or, in the Babylonian tradition, Nergal.[15] Notably, the tradition of

---

thunderbolts," see my *The Cosmic War,* pp. 28–66.

9 Mayor, *The First Fossil Hunters,* pp. 195–196.

10 Mayor, *The First Fossil Hunters,* p. 196, emphasis added.

11 Ibid., p. 265. Mayor cites Josephus' *Antiquities of the Jews,* 5:23 in which he describes the giants as having "bodies so large and countenances so entirely different from other men, that they were amazing to the sight and terrible to the hearing." Note carefully that Josephus implies that giants are another *kind* of men.

12 Ibid., citing Manilius' *Astronomy,* 1:424–31. Manilius states that the giants were "broods of deformed creatures of unnatural face and shape" and that they were destroyed long ago during the time "when mountains were still being formed."

13 Ibid., p. 263.

14 Ibid., pp. 261–262.

15 Ibid., p. 263.

the "gods" annihilating the giants is dimly reflected in the biblical tradition, where Yahweh orders the extermination of the populations of Canaan during the Hebrew conquest, a population that according to the biblical text is of giant stature.[16]

The other interesting source is Herodotus, whose credentials as an accurate historian are, to say the least, a matter of hot debate within modern academic circles. In Herodotus' case, the example Mayor cites is of a coffin over ten feet long, in which was found a skeleton of a "man" as long as the coffin.[17] One may take this at face value, or rationalize that the "skeleton" was of some unknown creature whose bones had been arranged to *look* like a hominid, but really was not.

2. The Griffin

The latter strategy of rationalization is that followed by Mayor in a fascinating examination of the legendary creature, the griffin. Noting that the griffin played no real role in Greek mythology,[18] nonetheless there are artistic depictions of such creatures in Greek art. One of the most interesting aspects of accounts of griffins according to Mayor is both their consistency,[19] and the fact that they were consistently described as guarding "treasures."[20] Seeking a scientific explanation for this consistency and the origin of these legends, Mayor traces it back to the Gobi Desert tracer gold fields, and to the fact that the same fields exhibit fossils of the dinosaur protoceratops, which bears a strong resemblance to the griffin.[21]

3. Mayor's Interpretive Paradigm

The problem of the griffin highlights the same problem we encountered in chapter one — the sirrush — and the academic agendas in place to deal with such problems. For Mayor, the answer lies in the "mythological paradigm" itself:

> If some nonhuman features were detected in fossil assemblage, they could be explained by the mythological paradigm. Everyone knew that giants and heroes of myth were not merely bigger and stronger

16 See the statements of Numbers 13:33: "And there we saw the giants, the sons of Anak, [which come] of the giants: and we were in our own sight as grasshoppers, and so we were in their sight."

17 Mayor, *The First Fossil Hunters,* p. 264.

18 Ibid., p. 16.

19 Ibid., p. 34.

20 Ibid., p. 33.

21 Ibid., pp. 37, 41.

> than ordinary humans, but they could also have grotesque anatomical features, such as multiple heads or animal parts.[22]

Just how this paradigm works out practically can be seen in Mayor's other signal study of such mythologies, *Fossil Legends of the First Americans.*

B. Indians, Giants, Monsters, and War

1. *The Age and War of the Giants and Monsters*

The tradition of monsters, giants, and of a "war with the giants" is mirrored in the unlikely place of Native American Indian traditions and legends. As with the traditions of giants and gigantomachy in the eastern Mediterranean, Mayor notes that it is necessary, when viewing these Indian legends and traditions, to "keep in mind that in many Native traditions, 'giants' of ancient eras were often understood to be primeval beings that were neither animal nor human."[23] That said, this indigenous Native American tradition is very rich and diverse.

Like the early colonial Christians in North America who sought to interpret fossil evidence within the context of biblical stories of creation, the Nephilim, and the Flood,[24] Native American Indians, when encountering such bones and other evidence, "turned to mythic traditions about giants and monsters to account for them..."[25] The uniformity of this tradition of ancient giants and monsters across different tribes even called forth a comment from the famous Puritan Cotton Mather.[26] As for their Puritan counterparts, the Indians, like the Greeks, *interpreted such fossil evidence within the context of their already-existing tribal traditions about human prehistory.* The method in both cases is identical.

There are but two logical ways in which to view such traditions, be they biblical, Sumerian, or Native American, and they are that (1) either the myths were *created* by those cultures to explain such fossil evidence; *or* (2) the myths were handed *down* to such cultures and contained some kernel of actual historical truth, or to put it somewhat more provocatively, *the myths pre-existed both the evidence which was to be encountered and interpreted, and the cultures that would encounter and interpret it.* This is a phenomenon we have already encountered with the Greeks, and to a certain extent, it is true of *all* cultures of ancient times and their attempts to understand and interpret such evidence.

22 Mayor, *The First Fossil Hunters,* pp. 113–114.

23 Adrienne Mayor, *Fossil Legends of the First Americans,* p. 36.

24 Ibid., p. 33.

25 Ibid., p. 37.

26 Ibid., p. 36.

Needless to say, the first alternative is that favored by academia and may be designated "the standard view." As we proceed with our survey of Native American traditions, however, we shall see that there are a number of things that suggest that the second alternative, for all its radical nature, is the more rational alternative, and one deserving of a detailed exploration.

Just as for the Greeks across the Atlantic, the nature of the giants was not a settled matter for Native Americans. In some traditions, the giants were said to be made of stone and lived almost 1300 years before the arrival of Columbus, according to the Iroquois scholar David Cusick;[27] in others, the giants were "humanoid" creatures,[28] and there were debates within, and differences between, traditions over whether or not the giants were even hostile or indifferent to humans.[29]

There is an amazing consistency of Native American traditions regarding the "age of the giants and monsters" and in some traditions, the war that was fought against them. Again, according to the Iroquois scholar David Cusick, the "northern giants, called *Ronnongwetowanea,* had harassed the early Iroquois in the past, but the giants all died out about twenty-five hundred winters before Columbus discovered America."[30] Running these numbers (1492 - 2500 = 1008 B.C.) places the Native American account of the extinction of the giants and the end of the "age of giants" at *very* roughly the same period of time as biblical accounts of the Hebrew conquest of Canaan, in which giants were a specific target for extinction.[31] Other Iroquois tribes placed the death of the last giant at eight to ten generations prior to 1705,[32] again, in a time frame roughly consistent with other Native American traditions and broadly consistent with similar legends from the Middle East.

One of the most intriguing, and as we shall discover, most important, features of Native American traditions and legends concerning the giants and monsters is the fact that many of these traditions taught the idea that *various "past ages (were) distinguished by different kinds of creatures,"* a belief that was *"a long-standing concept in many Native American traditions,* and discoveries of unusual vertebrate fossils would certainly reinforce the idea."[33]

Among the Aztecs in Mexico, this idea found further expansion, and in

27 Adrienne Mayor, *Fossil Legends of the First Americans,* p. 41.

28 Ibid., p. 34.

29 Ibid., p. 36.

30 Ibid., p. 40.

31 There are of course the "early" and "late" dates for the Exodus, and hence, the Conquest, in standard biblical scholarship, but more recently a school of thought — the "New Chronology," led by its premier champion, David Rohl — has arisen to push the date of the Exodus some three centuries further back into history, ca. 1700 B.C.

32 Mayor, *Fossil Legends of the First Americans,* p. 34.

33 Mayor, *Fossil Legends of the First Americans,* p. 38, emphasis added.

the expansion, an eerie parallel with the Mesopotamian and Middle Eastern accounts suggestive of an engineered humanity:

> In Aztec mythology, there were four previous ages of the world, each destroyed by a different cataclysm: flood, earthquake, hurricane, and fire. *The first age was dominated by the earth-giants, followed by three ears of primitive humans. The Aztecs believed that inhabitants of the later worlds sometimes encountered terrifying giants who were relict survivors of the great flood and earthquakes that had destroyed the past worlds.* To re-create life in the present, fifth age, the Feathered Serpent god Quetzalcoatl retrieved the scattered and broken bones of the human ancestors destroyed in the fourth age. *He ground the bones to powder in a jade mortar. Mixed with blood donated by the gods, these bones produced today's humans.*[34]

There are a number of very important points to note here.

Firstly, the Aztec tradition is broadly consistent with North American Native American traditions of different ages distinguished by different creatures, which suggests three possibilities for the resemblance: (1) either both traditions stem from a common and earlier source; or (2) all Native American cultures were in much closer contact with each other than the Isolationist school championed by the Smithsonian and "official archaeology and anthropology" would have it; or (3) some combination of (1) and (2) was true. It should be noted that if the Isolationist interpretation were true, then option (1) would be a way to explain it, but this would present academia's "standard view" with the problem of having to explain why so many disparate tribal traditions maintained the concept with such consistency over a wide area and prolonged period, and "independently" of one another. Conversely, if the disparate traditions did *not* spring from a common source, then how would one account for the amazing similarity of "the mythological imagination" over such a wide area — a similarity, moreover, that bears amazing resemblances to modern evolutionary theory? In other words, the consistency of the concept itself strongly suggests a scientific basis *from which* the various mythologies arose, and thus suggests a time and culture antedating the Meso- and North American Native traditions and stemming from very high antiquity.

Secondly, within the Aztec tradition, explicit mention is made of different *types* of pre-existing humanity, a conception well in keeping with modern evolutionary theory concerning the origins of modern *Homo sapiens sapiens*, with one *very* important exception, and that is that modern mankind in the Aztec

34 Ibid., pp. 89–90, emphasis added.

tradition, as in the Mesopotamian, is an *engineered* creature, and one moreover that is chimerical, i.e., composed of a part from "the gods" and a part from the pre-existing and more primitive "humans." Even the details of mankind's creation are eerily parallel with the Kharsag tablets and the interpretation of the O'Briens examined previously, for we have (1) a grinding of the bones to a powder, paralleling the O'Briens' creation of a culture, and (2) the "donation" of the blood — meaning perhaps the semen emissions — of the "gods" to the hybridized creature. Again, *the details of the Aztec creation myth suggestively point to a technological basis and to an engineered, and* ***not*** *to an evolved humanity.*

Finally, in yet another odd parallel with the Middle East, the Aztecs believed that *some* of the giants survived the Flood. In this, there is also general alignment with the traditions of more northern Native American nations which believed that the wars with the giants were a post-flood event.

The tradition extended beyond the Iroquois in North America or the Aztecs in Mexico. For example, as Cortez and his men marched westward into Mexico toward the Aztec capital Tenochtitlan, they encountered the tribe of the Tlaxcaltecs or Tlascala Indians, who recounted for Bernal Diaz del Castillo, one of Cortez's captains, that

> A very long time ago, their forefathers found the territory inhabited "by men and women of great size, people with huge bones." The ancestors had fought and destroyed these "wicked and evil" beings — and "any of the giants who survived eventually died out." This last detail reveals that the Tlaxcaltecas understood that even if a small number of relict creatures had escaped mass destruction, they would eventually face extinction.[35]

Again, the legend compels the observation that the Tlaxcalteca account is remarkably *parallel* with the biblical account of the conquest of Canaan by the Hebrews, in that both peoples (1) encounter giant humanoid occupants of the land, and (2) wage war against them because these giants are "wicked and evil." This observation compels three further questions: Are we looking at a "conquest" that occurred in more than one place but for the same reasons? *If so, then we are probably looking at coordinated action and an agenda, namely, a genocidal war for the extinction of a certain race or species of "giants."*

Or alternatively, are we looking at dim memories in either case of one underlying event that occurred in the dim mists of "pre-history," or are we looking at some combination of both? If the latter two cases be true, then this in turn would perhaps have a wide and profound impact on how the editing

35 Mayor, *Fossil Legends of the First Americans,* pp. 74–75.

of biblical and other Middle Eastern texts is understood to have occurred, and *might* even suggest possible reasons for why it was undertaken.

Note also that the tradition here is clear: the occupants that the Tlazcalteca encountered were *living human-like creatures of large stature.* They were *not* merely fossilized bones that were interpreted in accordance with a pre-existing myth. This will become an important point in a moment.

The Aztecs added to this "giant lore" when, during their migrations into lower Mexico, they encountered ca. 1200 A.D. the abandoned city of Teotihuacan, the famous giant pyramid complex outside of modern Mexico City. Seeing these gigantic structures, they interpreted them as having been built by the giants during the age of the giants.[36] The Aztec prince Fernando de Alba Ixtlilxochitl maintained that these giants were "earth-giants," in a manner recalling the far-distant Iroquois' "stone giants," and that they were somehow deformed.[37]

Even farther south in Latin America, the Incas had similar traditions of giants, monsters, and wars, and explained "colossal skeletons as the vestiges of dangerous giants of antiquity."[38] Cieza de Leon conducted interviews with local Manta Indians in Ecuador, and they had traditions that "Had been received from their ancestors from very remote times"[39] that a race of giants had arrived on the coast of such stature that ordinary men came up only to the knees.[40]

Here again one encounters a story with odd and out-of-place resemblances to yet another biblical story, from an entirely different tradition an ocean away. De Leon, in his recounting of the Native traditions, insisted

> that because of their vile sexual habits, the giants were "detested by the natives," who made war against the invaders in vain. At last God intervened, and while the giants "were all together engaged in their accursed [words omitted], a fearsome and terrible fire came down from heaven with a great noise. At one blow, they were all killed, and the fire consumed them."[41]

While one is left guessing what the "vile sexual habits" might be, the resemblance to the biblical story of the destruction of Sodom and Gomorrah is quite strong, and this suggests a rather unique twist to the latter, for if both traditions come from some common underlying source and represent fragments

---

36 Mayor, *Fossil Legends of the First Americans,* p. 88.
37 Ibid.
38 Ibid., p. 80.
39 Ibid.
40 Ibid.
41 Ibid.

of a once-unified story or legend, then perhaps the destruction of Sodom and Gomorrah has less to do with the conventional religious and moral explanations and more to do with the presence of giants, or conversely, perhaps the destruction of the giants in the Americas had something to do with the morality of Sodom and Gomorrah.

2. The Consistency of Native American Explanations

What emerges from all this is the amazing consistency of Native American traditions concerning the age of giants and monsters, for with the exception of the Tlaxcaltecas noted previously, "Native American traditions about giants and immense land animals indicated that such creatures had never been seen alive in the present age, that the huge beasts had disappeared long ago..."[42] Moreover, "The Natives' story was consistent: (fossilized bones) were the vestiges of a giant race, now extinct owing to natural catastrophe or battles with humans of the distant past."[43] In this respect, Native Americans had a tradition that interpreted the evidence they encountered in a manner remarkably similar to the Greeks, who had their *own* traditions of giants, monsters, and ancient wars against both.

3. Mayor's Explanation

Mayor's scholarly credentials and the caliber of her research are impeccable. Not surprisingly, she follows a standard academic interpretation of the Native American traditions. Citing again the case of the Tlaxcaltecas who recounted tales of giants pulling down trees "as if they had beene stalkes [sic] of lettices [sic]," she notes that such behavior is similar to that of elephants, and that such details "may have originated in ancestral memories of Columbian mammoths and may have been later confirmed by discoveries of fossils."[44] Similarly, all such tales of giants' remains might simply be explicable as misinterpreted fossilized bones of "monsters" or dinosaurs.[45] For Mayor, the bottom line seems to be that *the myths were created to explain the evidence.*

But the problem, as we saw — at least in the case of the Tlaxacaltecas (if not others) — was that the legend explicitly stated that their ancestors first encountered *humanoid* giants as living creatures, and the Tlaxcaltecas were no less rational than anyone else, and would hardly be prone to confuse humanoid beings with mammoths.

42 Ibid., p. 79.
43 Ibid., p. 78.
44 Ibid., p. 77.
45 Ibid., p. 78.

The other problem is that if the myths were created to explain the evidence, then why did cultures as diverse as the Greeks, Iroquois, Aztecs and so on, creates myths remarkably similar in their details?

4. The Ancient Traditions and the Alternative Explanation

Thus, there are certain factors that suggest that an alternative view of these traditions, myths, and legends is necessary, one in which it is assumed that they contain kernels of historical and scientific truth:

1) The belief among some Indian tribes and nations that different *ages* were exhibited and inhabited by different types of *creatures, including different types of humanity;*
2) The consistency with which Native Americans pointed to fossils as evidence of the truth of their traditions and myths concerning the age of giants and monsters (a fact that, again, would indicate that the myths predate the cultures encountering the evidence);
3) The belief of different ages being typified by different kinds of creatures and humans, coupled with the consistency of their understanding of fossils as coming from antiquity and a war of giants and monsters, suggests not that the Indians invented the myths to explain the bones, but rather *that the origin of those myths and traditions stems from the time of the dinosaurs themselves, for if such myths contained any kernel of truth, then at some point they had to be based on contemporaneous observation.*

Why is this so? Why must one entertain the possibility that these myths and traditions stem *from the time of the dinosaurs themselves*?

Look carefully at point number one above, for such a viewpoint is remarkably similar to the ideas of modern evolutionary theory and modern anthropology. Indeed, there is little if anything to distinguish the American Indian views, when reduced to the barest elements, from the views of modern science. The legends and myths, in other words, suggest an origin within a culture far more scientifically sophisticated than those Indian tribes which preserved them, and that means in turn that the origin of these traditions is very, very ancient, or, to put it in the terms I have used elsewhere, "paleoancient." Indeed, if they originate from such a culture and more or less contemporaneously with the events described, *then they antedate modern mankind himself.*

To put it as succinctly and nakedly as possible, *the Native American Indian traditions are older than modern mankind himself and thus predate the tribes that preserved them. And this act must be considered to be an act of "preservation" and not "creation" simply because the scientific sophistication they suggest could not have originated within those tribes and cultures.*[46]

This observation, plus the observations of all previous chapters, now suggests that two stunning and mutually opposed *agendas* might be in play "from high antiquity":

1) On the one hand, there is a body of lore, myths, traditions, and legends, spanning the globe from ancient Mesopotamia to North and Meso-America, which, taken together, suggest that modern mankind is the deliberately engineered product of some genetic "cousins" who were here long before modern man emerged. Additionally, both Mesopotamian, ancient Hellenic, *and* Native American Indian traditions speak of a war with "giants and monsters" occurring at some point in "high antiquity" prior to the emergence of modern man, and yet with whom at some later point modern man is contemporaneous. In some Native American Indian traditions, this view is codified into the belief that there were different ages of humanity and that these ages were also typified by different types of creatures that populated them. *Thus, one agenda suggested by these observations is that someone from high antiquity wished modern mankind to know his true origins, and how these fit into a larger picture involving wars, giants, and "monsters."* Furthermore, the presence within ancient texts of details suggestive of an ancient high technology of genetic engineering also suggests that the "monsters" themselves — i.e., the chimeras of myth *and* the dinosaurs of science — might themselves be the deliberate products of engineering.
2) On the other hand, there is a body of lore, myths, traditions and legends embodied in *some texts* that suggest that others, at a later period, wished to *obscure and hide* those origins from a segment of humanity, and did so via the technique of "religifying" them in ancient times, and in modern times, ruling such possibilities out of court in the name of "science." In the service of this *possible* agenda, ancient texts and legends have *perhaps* been deliberately

46 With respect to this observation it is also important to recall that many of these Native American nations also had legends and traditions that explicitly stated that they originated "elsewhere," and in some cases these legends also imply an origin entirely off this planet. Other Native American traditions from Meso- and South America also explicitly state that their cultures are *legacy* cultures and the product of some "civilizer god" from whom they derived their scientific and cosmological sophistication (see my *The Cosmic War*, pp. 279–285).

> mistranslated to obscure the possible technological references. Viewed a certain way and from this perspective, both classical Judeo-Christian texts and modern evolutionary dogma each serve to obscure these origins *if* one grants the proposition that there is any truth to such ancient myths that suggest that humanity is an engineered product, and that it had "cousins out there" that *did* the engineering. The very *fragmentation* of the story into so many disparate traditions with conflicting details over the basic storyline also suggests that the fragmentation itself may have been an attempt to obscure the story.

The astute reader will now have noticed an acute problem: if the legends concerning giants and monsters are true on the one hand, and if they are to be taken as indicative of something peculiar going on in the age of the dinosaurs on the other (as one implication of the standard academic view would have it), then *who* is doing the "remembering" here and creating the myths to begin with? If some Indian traditions are sophisticated enough to suggest different ages *populated by different creatures, including different "humanities,"* then this is a view every bit as sophisticated and "scientific" as modern theories of the origins and evolution of mankind.

But clearly, sophisticated as these Native American traditions and cultures were, they were not sophisticated enough to have such a scientific view.

In short, they did not create their myths, they *inherited* them.

So once again, who *is* creating these stories, and why?

The answer to that question requires, once again, a foray into current scientific and genetic findings on the origins of man, and a careful combination of that science with the views of the ancient texts that I have advanced here and elsewhere...

Eight

A Memory of Man Past:

Genetic Clans, Archaeological Anomalies, Evolutionary Enigmas, and Speculative Solutions

∴

"Thus 'facts' turn out to be networks of arguments and observational claims."
—Michael A. Cremo and Richard L. Thompson[1]

EUROPE IS POPULATED by approximately six hundred and fifty million Caucasians, as of this writing. No news there. The news is that all of them, along with their American, Canadian, and Australian descendents across the seas, are "cousins." The reason is that Bryan Sykes, a geneticist at the University of Oxford,[2] discovered an amazing thing: all Europeans come from only seven different clan mothers, or *The Seven Daughters of Eve* as he calls them, which is the title of his fascinating book surveying his research and conclusions. Those seven different clan mothers all in turn come from one common mother, "mitochondrial Eve" as the geneticists call her, as indeed, do all humans now alive on the planet.

If this sounds like good news to the biblical literalist, it isn't, but we'll get to that later. The real news, here as elsewhere, is in the genes, and what they tell us about human origins and prehistory. Sykes, with much more eloquence and elegance than most scientists, puts it this way:

1 Michael A. Cremo and Richard L. Thompson, *Forbidden Archeology: The Hidden History of the Human Race*, p. 19.

2 Bryan Sykes, *The Seven Daughters of Eve: The Science that Reveals Our Genetic Ancestry* (W.W. Norton & Co., 2002), p. 10: Sykes notes that he is a professor of Genetics and the Institute of Molecular Medicine of the University of Oxford. Sykes' book is an excellent and enjoyable read, and an essential component for anyone wishing to examine the claims of ancient texts concerning human origins.

> ...(Each) of us carries a message from our ancestors in every cell of our body. It is in our DNA, the genetic material that is handed down from generation to generation. Within the DNA is written not only our histories as individuals but the whole history of the human race. With the aid of recent advances in genetic technology, this history is now being revealed. We are at last able to begin to decipher the messages from the past. Our DNA does not fade like an ancient parchment; it does not rust in the ground like the sword of a warrior long dead. It is not eroded by wind or rain, nor reduced to ruin by fire and earthquake. It is the traveller from an ancient land who lives within us all.[3]

Sykes' book is the record of his research efforts to trace that history back into "the deep past" of high antiquity, utterly "beyond the reach of written record or stone inscriptions."[4] That research revealed that Caucasian Europeans and their descendents were traceable to "only a handful of women living tens of thousands of years ago."[5]

A. Mitochondrial Eve and her Seven European Daughters

1. Early Attempts to Distinguish Racial Groups by Blood-Typing

This "handful of women" were in fact seven women, seven "genetic clan mothers" to whom Sykes gave the names Ursula, Xenia, Helena, Velda, Tara, Katrine, and Jasmine.[6] But the quest for science to find and identify these seven genetic clan mothers of Europe was no easy matter.

It began with the first blood transfusions in Italy in the seventeenth century.[7] Many people died from reactions to these early attempts and the practice was discontinued for two more centuries until it was resumed again in the nineteenth century to try to save the lives of women hemorrhaging during childbirth. It was the practice of transfusions, in other words, that led to the discovery of the different blood types of humanity by the biologist Karl Landsteiner in 1900.[8]

Blood-typing quickly became an acceptable way of determining paternity and, ultimately, of the attempt to trace the genetic and evolutionary origins of different groups of humanity. According to Sykes, the attempt began in

---

3 Sykes, *The Seven Daughters of Eve,* p. 1.
4 Ibid., p. 2.
5 Ibid.
6 Ibid., pp. 8–9.
7 Ibid., p. 33.
8 Ibid., p. 34.

earnest during World War I, and "to a scientific paper delivered to the Salonika Medical Society on 5 June 1918."[9]

Because the genetics of inherited blood types was rather straightforward, the attempt to classify racial groups by blood types was soon underway. On the basis of the transfusion data collected during World War I, it was soon discovered that Europeans were made up of about 15 percent blood type B and 40 percent blood type A. But these proportions changed the further East into Russia that one went, where blood type B was proportionally higher in soldiers drawn from Russia and Africa, with the proportion peaking to about 50 percent in soldiers from India serving with the British. This led some researchers to conclude that there were two basic or early "bio-chemical racial groups," A and B, based on blood typing.[10]

But problems with this method quickly arose. For example, one result showed almost identical blood type frequencies occurring in soldiers from Russia as soldiers from Madagascar! As Sykes quips, was this "genetic evidence for a hitherto unrecorded Russian invasion of Madagascar, or even the reverse, an overwhelming Malagasy colonization of Russia?"[11] Other results showed frequencies nearly identical between the English and the Greeks. These types of problems grew so acute that the American physician William Boyd eventually issued a warning to anthropologists to disregard blood types as any reliable indicator of the genetic history of humanity and its different races.[12]

2. The Basques, Rh Positive and Rh Negative Blood, and the "Problem of Europe"

At this juncture, there was another genetic monkey wrench thrown into the works: the Basques. I have always been fascinated by the Basques, because, being part Basque myself on my mother's side of the family, their strange relationship to the rest of Europe is somehow part of my own personal ancestry and story. The Basques inhabit the area of the now long-defunct Kingdom of Navarre in the corner of the Bay of Biscay in the area where the modern borders of France and Spain touch.

The problem for anthropology that they pose is twofold, for on the one hand, the Basques are the European continent's "most influential genetic population," and on the other, their language, Euskara, "is unique in Europe in that it has no linguistic connection with any other living language."[13]

9 Sykes, *The Seven Daughters of Eve,* p. 34.

10 Ibid., p. 36.

11 Ibid., p. 37.

12 Ibid.

13 Sykes, *The Seven Daughters of Eve,* p. 39.

But they also provide Sykes and his research team "with an invaluable clue to the genetic history of the whole of Europe..."[14]

The clue comes through the different Rh positive and Rh negative blood types. Most people are aware of the severe complications that can occur for a newborn baby born of an Rh positive and an Rh negative pair of parents. "Blue baby syndrome" was a common occurrence of births for European peoples until this distinction was discovered, and Rh negative mothers married to Rh positive husbands were given injections of antigens that neutralized the mother's immune system reaction to it and hence protected her child from accidental circulations of both kinds of blood in her baby.[15]

The problem was that while most of the rest of the world was overwhelmingly Rh positive, in Europe alone there was a nearly equal mixture of both Rh positive and Rh negative types. And this "did not make any evolutionary sense."[16]

It is at this juncture that the Basques assumed a crucial role in the story, for in 1947 the English physician Arthur Mourant decided to study the problem posed by the Basques more closely. The results were somewhat astonishing, as Sykes explains:

> It was already known that Basques had by far the lowest frequency of blood group B of all the population groups in Europe. Could they be the ancient reservoir of (Rh) negative as well? In 1947 Mourant arranged to meet with two Basques who were in London attempting to form a provisional government and were keen to support any attempts to prove their genetic uniqueness. Like most Basques, they were supporters of the French Resistance and totally opposed to the fascist Franco regime in Spain. Both men provided blood samples and both were (Rh) negative. Through these contacts, Mourant typed a panel of French and Spanish Basques who turned out, as he had hoped, to have a very high frequency of (Rh) negatives, in fact, the highest in the world. Mourant concluded from this that the Basques were descended from the original inhabitants of Europe, whereas all other Europeans were a mixture of originals and more recent arrivals, which he thought were the first farmers from the Near East.
>
> From that moment, the Basques assumed the status of the population against which all ideas about European genetic prehistory were to be — and to a large extent still are — judged. The fact that they

---

14 Ibid. Sykes' more technical explanation is worth reading in detail.

15 Ibid., pp. 39–40.

16 Sykes, *The Seven Daughters of Eve,* p. 40.

> alone of all the west Europeans spoke a language which was unique in Europe, and did not belong to the Indo-European family which embraces all other languages of western Europe, only enhanced their special position.[17]

In other words, of all the population groups in Europe, the evidence appeared quite strong that the Basques were somehow "original" to the continent, or, better put, the group that had been there the longest and had arrived before the others.

The next step forward came, of course, with the discovery of the double helix structure of DNA itself, and with the technologies associated with genetic sequencing. Here at last was a technique that would allow scientists to stare down the long spiral and peer into the histories of various human groups. By comparing massive amounts of DNA and statistically quantifying certain clusters or sequences in the DNA, geneticists could derive an idea of the "genetic distance" between groups. The farther apart two groups were genetically, the more distant in the past any common ancestry was likely to be.[18]

Doing so, however, threw yet another monkey wrench into the works, for over and over again, in different racial groups, individuals would appear in one group whose closest genetic relatives were in an entirely different group.[19] Genetics, in other words, had blurred the traditional anthropological classifications based on race, and yet was also showing the emergence of distinctive groups *within* races such as Caucasians. Nonetheless, the concept of "genetic distance" *did* lead to one very interesting conclusion when *all* human groups were considered, for it meant that "the whole of the human race was much younger and more closely related than many people thought."[20] In fact, it meant that modern *Homo sapiens sapiens* has only been around for approximately the last 150,000 years![21]

Here we encounter the most significant problem of them all, and Sykes zeros in on it with his customary eloquence, for what was the relationship *genetically* between modern *Homo sapiens sapiens* and the earlier precursor species assembled from fossil records by paleontologists?

> Their names — *Homo habilis, Homo erectus, Homo heidelbergensis, Homo neanderthalensis* — reflect the to and fro of the attempts to pigeon-hole them into different species. However, these are species

17 Ibid., pp. 41–42.
18 Sykes, *The Seven Daughters of Eve,* p. 43.
19 Ibid., pp. 48–49.
20 Ibid., p. 49.
21 Ibid.

> defined on the basis of anatomical features preserved in skeletons, particularly the skulls, and not in the biological sense of different, genetically isolated, species who are incapable of breeding with any other. It is an operational classification with no evolutionary consequences. From the shapes of the bones alone there is simply no way of knowing whether humans (I use the term 'human' to include everything in the genus *Homo*) from different parts of the world were capable of successful interbreeding. If they could interbreed, then this opens up the possibility of their exchanging genes and spreading mutations around....
>
> It is this question that lies behind one of the longest-running and most deep-seated controversies in human evolution. Are the different species defined by paleontologists — *Homo erectus, Homo neanderthalensis* and ourselves, *Homo sapiens* — all part of the same gene pool or not? Or, to put it another way, are modern humans directly descended from the fossils found in their part of the world, or are many of these the remains of now extinct genetically separate human species?[22]

Bear in mind that point about the paleontological classification of different *species* within the genus *Homo*, for it will become very important later in this chapter.

These paleontological classification schemes emerged, as Sykes has indicated, by careful comparison of fossilized remains, most of which come from Africa. This important point led paleontologists and anthropologists to propose an origin for modern man "out of Africa," yet the presence of such remains in other parts of the world have led to a long-running debate. On the one hand, there are those who propose that modern *Homo sapiens sapiens* migrated out of Africa some 100,000 years ago. The opposite school, on the other hand, proposed a kind of "regionalism" wherein the species evolved, more or less simultaneously and spontaneously, in different parts of the world for similar reasons.[23]

But for Sykes, the fossil record, "incomplete and patchy though it is, consistently points to Africa as the ultimate origin of all humans."[24] And if the species paleontologists had classified on the basis of that fossil record *were* in the evolutionary phylogenetic tree of modern man — that is, if modern man evolved from these other species — then could genetics resolve the debates?

22 Sykes, *The Seven Daughters of Eve,* p. 110.

23 Ibid., p. 50.

24 Ibid., p. 111.

Was there, for example, any evidence that Neanderthal man and Cro-Magnon man had any genetic commonality, and thus possibly some deeper common origin?[25]

Indeed it could.

When mitochondrial DNA from Neanderthal remains were sequenced and compared to that of approximately six thousand modern Europeans, it led to the conclusion that modern man and Neanderthal man could not be related any later than a quarter of a million years ago. Indeed, the sequencing led to the conclusion that not only were modern Europeans *not* survivors of Neanderthal man, Neanderthal was not an ancestor.[26] There was a complete *absence* of Neanderthal mitochondrial DNA in modern Europeans.[27]

But that wasn't the only story that mitochondrial DNA told...

3. Mitochondrial DNA and the Y Chromosome: Mitochondrial "Eve" and Chromosomic "Adam"

a. Mitochondrial DNA and the Seven Clans

...that story, the story of "mitochondrial Eve" and her seven European "daughters," all clan mothers to all Europeans, told quite a story, but to see how, we must know a bit more about genetics.

All genes are inherited from both sets of parents, with but two exceptions: (1) mitochondrial DNA, and (2) the male Y chromosome. Within the mitochondria of every cell of every animal, including humans, there is a small amount of "mitochondrial DNA," which in the case of humans comprises a mere 16,000 base pairs out of the three *billion* pairs in human DNA. Surprisingly, this mitochondrial DNA is also coiled in upon itself in a circle.[28] What is unique about mitochondrial DNA is that each human receives it *only* from his or her mother. The reason is that when the male sperm fertilizes an egg, the sperm's mitochondrial DNA is ejected along with its tail, and only the sperm's nuclear DNA is paired with the mother's DNA; the mother's mitochondrial DNA, however, enters every mitochondrion of the new human being.[29] Thus, while both men and women possess the mitochondrial DNA of their mothers, only *women* pass it on to their offspring.

We may pause here and note that among the ancient Egyptians and Hebrews, lineage was always traced through the mother, a legacy, perhaps, of a scientific culture which preceded them.

---

25 Sykes, *The Seven Daughters of Eve,* p. 117.
26 Ibid., p. 126.
27 Ibid., p. 127.
28 Ibid., p. 53.
29 Sykes, *The Seven Daughters of Eve,* p. 54.

But this is only half the story. Mitochondrial DNA also mutates much more quickly than ordinary DNA, and thus, "the 'molecular clock' by which we can calculate the passage of time through DNA is ticking much faster" within mitochondrial DNA versus nuclear DNA.[30] This allows geneticists to calculate the approximate times or periods when significant divergences within "clans" of mitochrondrial DNA emerged.

But there's still more...

When European mitochondrial DNA was sequenced and appropriate methods of applied mathematics were developed to determine the relative "clustering" of these results,[31] the result was rather astonishing, and here, once again, the Basques enter the story, and again upset the applecart of what had been assumed up to that point. Prior to the development of sequencing techniques and the appropriate mathematical models, it had been assumed that an agricultural explosion in the Middle East had led to a gradual migration of people from that region into Europe, gradually replacing the sparser hunter-gatherer population that had been assumed to exist in Europe. But Sykes and his team found that only one of the seven "mitochondrial DNA clusters" that they had found fit that description.[32]

Sykes and his team were not initially convinced to abandon the standard view, until they once again checked with the sequencing results of the Basques. For the standard view to be true, the Basques should have shown up to be a unique group within the rest of Europe. But just exactly the *opposite* was the case. The Basques turned out to contain representatives from six of the seven European mitochondrial DNA clusters.[33] Most Europeans, in other words, came from six clan mothers "indigenous" to Europe, while the seventh group came into the picture at a later point.

b. The Y Chromosome and the "Ten Fathers"

There was another story that was being told by genetic sequencing as well, and that story is locked up in the male Y chromosome. The Y chromosome has, as Sykes quips, but "one purpose in life: to create men."[34] In effect, it is a stunted gene which *prevents* human embryos from becoming girls.[35] It literally programs other genes in human DNA to develop into males and not females.

30 Ibid., p. 55.
31 See Sykes' discussions on pp. 137–141.
32 Ibid., pp. 142–143.
33 Sykes, *The Seven Daughters of Eve,* p. 143.
34 Ibid., p. 187.
35 Ibid.

In other words, it is a "special program."[36]

Its presence in males meant that a similar technique for tracing the paternal ancestry of men could be developed as for tracing everyone's ancestry through mitochondrial DNA, with one very significant exception: it could only be done for men because — with a minor exception — it was found only in men.[37] The problem with the Y chromosome, however, proved to be its extraordinary stability; they were not only "full of 'junk' DNA which had no obvious function,"[38] but all over the world the amount of mutations was far below what was expected. Fortunately, however, there were "repeats" in short segments of the Y chromosome that allowed proper genetic fingerprinting of it to be done. When this was done on European and Middle Eastern males, it led to the identification of ten clusters, or, "clan fathers."[39]

The bottom line is this: every European human comes from one of seven clan mothers, and every European male comes from one of ten clan fathers.

But there's more...

4. The Seven Mothers of Europe and Their Clans

The seven mitochondrial DNA clusters identified by Sykes and his team allowed them to determine that all seven had emerged between 45,000 and 10,000 years ago.[40] Determining their probable origin and migrations, however, was more difficult. For example, a clan predominant in Scotland now, and which shows origins from approximately 20,000 years ago, could not have originated in Scotland for the very simple reason that it was covered in ice 20,000 years ago.[41] Sykes and his team, on the basis of their clustering findings and a process of reasoning, essentially modified the standard model.

a. Ursula's Clan

"Ursula" was born 45,000 years ago,[42] and her clan now constitutes approximately 11 percent of the European population. While spread all over

36 Sykes notes that in rare occurrences, one in twenty thousand women are born with the Y chromosome, which makes them on average taller, but also means that their ovaries and uterus do not properly develop, making them infertile. Analysis of the Y chromosome in these unfortunate people shows that it contains a mutation which prevents its proper functioning.

37 See pp. 188–189 for a fuller discussion.

38 Sykes, *The Seven Daughters of Eve,* p. 190.

39 Ibid., p. 193.

40 Ibid., p. 196.

41 Ibid., p. 200.

42 Sykes, *The Seven Daughters of Eve,* p. 202.

Europe, there are concentrations in western Britain and Scandinavia.[43]

b. Xenia's Clan

"Xenia" was born approximately 25,000 years ago,[44] and her maternal descendents constitute about six percent of the European population, with three branches dispersed from Eastern Europe all the way into France and Britain.[45]

c. Helena's Clan

"Helena's" clan represents almost 47 percent of the European population, reaching into every corner of the continent.[46] The clan is approximately 20,000 years old.[47]

d. Velda's Clan

"Velda" lived approximately 16,000 years ago, most likely in northern Spain,[48] and only about five percent of Europeans are her descendents. Surprisingly, many of them live in the very northern reaches of Norway and Finland.[49]

e. Tara's Clan

One of the most interesting clans is "Tara's," who probably lived in the hills of Tuscany in northwestern Italy some 17,000 years ago.[50] Her clan constitutes about nine percent of the European population, which is oddly concentrated along the Mediterranean, the western edges of Europe, and, of all places, Ireland.[51]

f. Katrine's Clan

Another interesting clan is "Katrine's," who according to Sykes lived in the vicinity of modern Venice approximately 15,000 years ago. While only five

---

43 Ibid., p. 212.
44 Ibid., p. 213.
45 Ibid., p. 220.
46 Ibid., p. 233.
47 Ibid., p. 221.
48 Ibid., p. 234.
49 Ibid., p. 232.
50 Sykes, *The Seven Daughters of Eve,* p. 243.
51 Ibid., p. 251.

percent of modern Europeans are her descendents, they remain concentrated in the Mediterranean but her descendents can be found all over Europe.[52]

g. Jasmine's Clan

"Jasmine" was the last and latest clan mother, living after the end of the last Ice Age in a permanent settlement.[53] Like "Tara's" and "Katrine's" clans, her clan is concentrated in specific areas, having moved through the Iberian peninsula, ultimately ending in western Wales, Cornwall, and western Scotland, while another branch funnels into central Europe. Her clan constitutes about 17 percent of the European population.[54]

h. The Deeper Ancestry, the Beginnings of a Problem and Some Beginning Speculations

The seven European daughters of Eve pointed, however, to a much deeper ancestry, and it is here that we begin to encounter the themes of previous chapters, including that of ancient genetic engineering, and the problems for the standard views that they pose. The problem arises because Sykes and his team then took the same methods of analysis and reapplied them to the seven European clan mothers themselves, and came to the astonishing conclusion, based on mitochondrial DNA mutations,[55] that they in turn had a common ancestor who most likely lived in the Middle East and long before humans settled Europe in significant numbers.[56] Sykes comments that it is "through this woman, [that] the whole of Europe is joined to the rest of the world."[57]

Tracing mitochondrial DNA sequences even further back, geneticists have been able to determine that there are *33* clans present in the world, an odd number, since, of course, it corresponds to the 33 degrees of Scottish rite Masonry. And of these, *13* clans are from Africa.[58] While the odd correspondence between the number of genetic clans and Masonry's "sacred number"

---

52 Ibid., p. 259.

53 Ibid., pp. 260–261.

54 Ibid., pp, 269–270.

55 Sykes, *The Seven Daughters of Eve,* p. 273.

56 Ibid., p. 274. The Middle Eastern origin of European populations raises its own special issues for alternative researchers, since the legends indicating a possible genetically engineered origin of humanity stem, in part, from there, and it is these legends that contain the greatest amount of detail suggestive of such engineering. The problem is complicated and confounded by indications of connections between ancient Sumer and the Indo-Aryan cultures of India... but *that* relationship is, perhaps, the subject of another book and investigation.

57 Ibid., p. 274.

58 Ibid., p. 276.

of 33 may be and probably is coincidental, the occurrence of *13* African clans is intriguing, and we shall return to it presently.

Ultimately, of course, geneticists have traced all human origins to a common mother, "mitochondrial Eve," who lived in Africa some 150,000 years ago.[59] While some may leap on this fact as yet another "scientific confirmation of the Bible," it really is not, for all that such genetic evidence indicates is that of all possible ultimate clans, only "Eve's" survived. There *may* have been other such ultimate clan mothers but their stock died out for whatever reason.[60] Mitochondrial Eve's descendents spread from Africa up through the Middle East, and thence to colonize the rest of the world.[61]

A comparison of these results with the O'Briens' examination of the indications of genetic engineering in the Kharsag tablets is illuminating, and highlights a number of problems, for on the one hand, the results of modern genetics appear to diametrically contradict the O'Briens' analysis, but on the other, there are minor indications that confirm it.

There are strong arguments against the O'Briens' analysis presented by modern genetics, and these may be boiled down to essentially two points:

1) The age of modern mankind indicated by genetics is approximately 150,000 years old, yet the Kharsag tablets were written much later, and the impression given by the O'Briens' analysis is that the genetic engineering subtly indicated in them does not occur at that period of time. In short, the *timing* is wrong; and,
2) The location of the engineering, as the O'Briens understood it, was in the Middle East near modern-day Lebanon, whereas the genetics indicates an origin from Africa. In short, the *location* indicated in the tablets is wrong.

This augurs poorly for a correspondence between modern science and ancient texts.

But there are *very slight and minor* indications of some correspondence between these genetic findings and the Kharsag Tablets as the O'Briens interpreted them, and oddly enough, these lie in the *numbers* recorded in both:

1) In the Kharsag Tablets, the original number of human "clan

59 Ibid., pp. 276–277.

60 Ibid.

61 Sykes, *The Seven Daughters of Eve,* pp. 277–278. Sykes notes that this colonization had to be through the Middle East and not over the straits of Gibraltar because the latter was a deep water channel, making it difficult for the colonization of Europe to take place by that route.

mothers" was *14,*[62] whereas genetic science indicates the number of primordial African clans to be *13*, a close agreement and one suggestive perhaps of some deep correspondence between the two data points in high antiquity;

2) The number of European "clan mothers" — seven — is a numerical component of the number *14*, again, a very *slight* indicator that perhaps there is some deeply rooted historical and scientific basis for the claims the O'Briens make concerning the Kharsag Tablets.

While such "evidence" does not permit us to draw *conclusions* one way or another about the possible correspondences between modern genetics and these ancient texts, they do highlight the nature of the problems of synthesizing the two that now confront us, and suggest some speculative solutions. But a synthetic and speculative resolution of this problem will require a closer look at the wider scientific context from which to view the problem, to which we now turn.

B. Evolutionary Chronology of the Origins of Man and the Chronological Problem

1. Evolutionary Chronology of Human Origins and Proto-Humans

While scientists are often divided over this or that detail of the evolutionary origins of humans, there is a general consensus over the chronological relationship of various operational classifications of other species within the human genus. There are numerous presentations of this standard model but for our purposes we will rely on the article "Human Evolution" in the online Internet encyclopedia Wikipedia for its relative accessibility. It is best to lay this out in an ordered numerical sequence, noting that we summarize only the major components, and not disputed intermediary species posited by some paleontologists and anthropologists:

1) Evidence from genetics suggests that 4 to 8 million years ago, gorillas and chimpanzees split from the line leading to humans. DNA evidence demonstrates that approximately 98 percent of human DNA is identical to that of chimpanzees;[63]
2) The first diversion within the origins of modern man according to the standard models occurs with the divergence of the genus

62 Christian and Barbara Joy O'Brien, *The Genius of the Few,* pp. 156–157; see chapter 5.

63 "Human Evolution," *Wikipedia,* www.en.wikipedia.org/wiki/Human_evolution, pp. 4–5.

Homo from Australopithecines approximately *2.3–2.4 million years ago*;[64]

3) *Homo habilis* flourished approximately 2.4–1.4 million years ago, evolving in southern and eastern Africa, diverging from Australopithecines;[65]
4) The next major evolutionary leap occurred with *Homo erectus*, who lived from approximately 1.8 million to a mere 70,000 years ago, the latter date indicating some contemporaneous existence with modern *Homo sapiens sapiens*. *Homo erectus* is believed also to have evolved larger brain capacity in some populations and to have fabricated and made use of simple stone tools, leading some to classify them as a separate species;[66]
5) There is still debate over where *Homo neanderthalensis*, which lived from approximately *400,000 years ago* (a significant date as we shall see), is a part of the evolutionary tree of modern man or no; that is, is it a separate species or is it a sub-species of *Homo sapiens*? DNA evidence suggests that the two species shared a common ancestor no later than 660,000 years ago. The *Wikipedia* article notes, however, that "a recent development in 2010 indicates that Neanderthal did indeed interbreed with Homo Sapiens at cerca [sic] 75,000 B.C. to create modern humans..."[67]giving a DNA content of modern humans that is approximately 1–4 percent Neanderthal, a significant amount given that humans and chimpanzees differ only in 1.5 percent of DNA.[68] What is *most* interesting, however, is that this 1–4 percent of DNA that is common to modern man and Neanderthal man is present only in non-African humans;[69]
6) Sometime between 400,000 and 200,000 years ago differences in skull cranial capacity developed, accompanied with a similar development in the sophistication of stone tools, allowing paleontologists to speculate these populations are the first beginnings of the genus *Homo sapiens;*[70]
7) Finally, as we have seen, the emergence of *Homo sapiens sapiens* occurred approximately 150,000 years ago, spreading from Africa, through the Middle East, to the rest of the world.

64 Ibid., p. 1.
65 Ibid., p. 5.
66 Ibid., p. 6.
67 "Human Evolution," *Wikipedia*, www.en.wikipedia.org/wiki/Human_evolution, p. 7.
68 Ibid.
69 Ibid.
70 Ibid., pp. 7–8.

Note what we now have in the major steps of the model summarized above, for the first *two* steps account for divergences of *genuses* within the primate family itself; in other words, we are not yet dealing with the genus *Homo* of "humanids" (if we may be permitted to coin that term), themselves.

It is with step 3 that we are dealing directly with evolutionary processes within the human genus itself, and note carefully what we have, for there are *essentially five major components or "steps" in that process.* If this sounds vaguely familiar, it should, for not only does it square in its broad outlines with the Native American Indian traditions of different ages typified by different humanities that were surveyed in the previous chapter, but more importantly, it will be recalled *that among the Aztecs there were exactly* ***four*** *such previous ages, giving a total of* ***five*** *"humanities" if one counts the current age and modern* Homo sapiens sapiens. [71] The case that the Native Americans *inherited* their legends from much higher antiquity has thus grown a bit stronger, for such observations imply by their very nature a more sophisticated pitch of civilization and scientific observation able to *make* such observations.

It is now time to throw the final monkey wrench into the works: the research of Vedic scholars Michael A. Cremo and Richard L. Thompson...

2. The Cremo-Thompson Archaeological Anomalies and Genetic Antiquity Problems

As noted, Cremo and Thompson are scholars of the Vedic literature of ancient India. But why would they become involved in a study of the scientific origins of man *in conjunction with* Vedic literature? Cremo's answer puts it "country simple":

> Some might question why we would put together a book like *The Hidden History of the Human Race*, unless we had some underlying purpose. Indeed, there is some underlying purpose.
>
> Richard Thompson and I are members of the Bhaktivedanta Institute, a branch of the International Society for Krishna Consciousness that studies the relationship between modern science and the worldview expressed in the Vedic literature of India. From the Vedic literature, we derive the idea that the human race is of great antiquity. For the purpose of conducting systematic research into the existing scientific literature on human antiquity, we expressed the Vedic idea in the form of a theory that *various humanlike and apelike beings have coexisted for long periods of time.*

71 Mayor, *Fossil Legends of the First American,* p. 38, see p. 234.

> That our theoretical outlook is derived from the Vedic literature should not disqualify it. Theory selection can come from many sources — a private inspiration, previous theories, a suggestion from a friend, a movie, and so on. What really matters is not a theory's source but its ability to account for observations.[72]

We share Cremo's and Thompson's convictions about the antiquity of humans — of whatever species — but adopt here as our methodology an even wider *context* in ancient texts, in what they indicate concerning human origins, where those origins come from, and, to a limited extent, the motivations recorded in those texts for the engineered creation of humans. We do so in order to propose a speculative resolution of the models of modern science on the one hand, and of the texts on the other.

Just how radical is Cremo's and Thompson's approach to this question may be readily appreciated by a brief survey of anomalous archaeological evidence that has been conveniently "forgotten" by modern science in its rush to avoid the implications that this evidence suggests. This evidence is presented in chapter six of their book *The Hidden History of the Human Race,* a chapter entitled, aptly enough, "Evidence for Advanced Culture in Distant Ages."[73] Here we survey a small but significant sampling of the data they have collected.

Dating by stratigraphy is a common method of dating the approximate age of an object embedded in certain geological layers of the earth... that is, unless the object and the stratum in which it is embedded is "inconvenient" to the standard model. One such object was reported in 1844 by Sir David Brewster, who stated that "a nail had been discovered firmly embedded in a block of sandstone from the Kingoodie (Mylnfield) Quarry in Scotland."[74] A nail embedded in sandstone would hardly be inconvenient, except in this instance the sandstone into which it was embedded happened to be from the Devonian period, making the date — if the find was genuine — between 360 and 408 million years old, *far* older than *any* evolutionary models for the origins of even the *genus* of "humanids," as shown above!

This wasn't all.

Cremo and Thompson note that on June 22, 1844, a story ran in the *London Times* that a gold thread (!) had been found embedded in a stone at a depth of eight feet. Once again, the stratigraphic context proved to be inconvenient, for the stone into which it was embedded was from

72 Michael A. Cremo and Richard L. Thompson, *The Hidden History of the Human Race* (Los Angeles: Bhaktivedanta Book Publishing, 1999), p. xix, emphasis added.

73 Ibid., pp. 103–122.

74 Ibid., p. 105.

the Early Carboniferous age, that is, "between 320 and 360 million years old."[75]

There was more bad news...

On June 5, 1852, an article innocently and innocuously entitled "A Relic of a Bygone Age" appeared in *Scientific American.* Cremo and Thompson cite the relevant portion of the article for our purpose:

> "A few days ago a powerful blast was made in the rock at Meeting House Hill, in Dorchester, a few rods south of Rev. Mr. Hall's meeting house. The blast threw out an immense mass of rock, some of the pieces weighing several tons, and scattered fragments in all directions. Among them was picked up a metallic vessel in two parts, rent asunder by the explosion. On putting the two parts together it formed a bell-shaped vessel, 4-1/2 inches high, 6-1/2 inches at the base, 2-1/2 inches at the top, and about an eighth of an inch in thickness. The body of this vessel resembles zinc in color, or a composition metal, in which there is a considerable portion of silver. On the side there are six figures or a flower, or bouquet, beautifully inlaid with pure silver. The chasing, carving, and inlaying are exquisitely done by the art of some cunning workman. This curious and unknown vessel was blown out of the solid pudding stone, fifteen feet below the surface. It is now in the possession of Mr. John Kettrell. Dr. J.V.C. Smith, who has recently travelled in the East, and examined hundreds of curious domestic utensils, and has drawings of them, has never seen anything resembling this. He has taken a drawing and accurate dimensions of it, to be submitted to the scientific. There is not [sic] doubt but that this curiosity was blown out of the rock, as above stated; but will Professor Agassiz, or some other scientific man please tell us how it came there? The matter is worthy of investigation, as there is no deception in the case.[76]

As Cremo and Thompson note, the pudding stone of the area in which the vessel was blasted loose from the rock, dates from the Precambrian period; that is, the stratigraphic dating of the object would make it over 600 million years old! Cremo and Thompson's concluding words about this discovery say it all:

75 Cremo and Thompson, *The Hidden History of the Human Race,* p. 106.

76 Cremo and Thompson, *The Hidden History of the Human Race,* pp. 106–107.

> By standard accounts, life was just beginning to form on this planet during the Precambrian [period]. But in the Dorchester vessel we have evidence indicating the presence of artistic metal workers in North America over 600 million years before Leif Ericson.[77]

And of course, metalworking art implies intelligence and technology to manufacture it.

It gets decidedly worse. A chalk ball was discovered in 1862 in strata 45–55 million years old near Laon, France, and reported in the April 1862 edition of *The Geologist.*[78]

A *coin*, with curious and indecipherable inscriptions, was discovered in Illinois in strata that would date it between 200,000 and 400,000 years ago (there's that date again!).[79] The finding was reported in 1871 by William E. Dubois of the Smithsonian Institution. What makes this particular coin so significant is what Dubois *said* about it, for noting its *uniform thickness,* Dubois concluded that it must have "passed through a rolling-mill; and if the ancient Indians had such a contrivance, it must have been pre-historic."[80] For Cremo and Thompson, the evidence again suggests "the existence of a civilization at least 200,000 years ago in North America."[81] And that, precisely, was the problem, for "beings intelligent enough to make and use coins (*Homo sapiens sapiens*)" were not around, according to the geneticists, until only 150,000 years ago![82]

There are clay figurines found at Nampa, Idaho in strata two million years old;[83] a gold *chain* embedded in Carboniferous coal discovered in Illinois in coal dating from 260–320 million years old;[84] an iron cup found in Oklahoma coal approximately 312 million years old;[85] a metallic machined rectangular *tube, in other words, a machined object,* discovered in France in layers of chalk 65 million years old;[86] and on and on the list could go.

One of the most interesting and most anomalous pieces of evidence recorded in Cremo's and Thompson's inventory of "inconvenient artifacts" is an obviously machined, small metallic sphere with three parallel grooves around its equator, found in strata from the Precambrian period in South Africa, said

---

77 Ibid., p. 107.
78 Ibid.
79 Ibid., p. 109.
80 Ibid., p. 110.
81 Ibid.
82 Cremo and Thompson, *The Hidden History of the Human Race,* p. 110.
83 Ibid., pp. 110–111.
84 Ibid., p. 113.
85 Ibid., pp. 114–115.
86 Ibid., p. 117.

to be almost three billion years old, almost three quarters of the age of the Earth itself![87]

All of this throws a rather thorny problem into the mix, for if, on the one hand, modern man *and his ancestors* are at best only 2.4 million years old, then *who was here doing all of this?* And note two dates here: the metallic tube discovered in France in chalk beds 65 million years old, and the Illinois coin that is between 200,000 and 400,000 years old, two significant dates as we shall see in a moment. But for now, the question is, if *Homo sapiens sapiens* is only 150,000 years old, and if it is generally agreed that he did not begin to use coins until only much later, then, once again, *who was here, and what were they doing?*

3. The Chronology of the Cosmic War and the Ancient Texts

If there are "inconvenient artifacts" tens, even hundreds, of millions of years old, or in one instance, *billions* of years old, and if these show evidence of machining and therefore of *technology*, then one answer immediately presents itself from the ancient texts: there were "others" here, our genetic cousins perhaps, who were clearly civilized, possessed of a technology, and perhaps every now and then, directly intervening in the course of human development and evolution.

And if this be the case, as Cremo's and Thompson's evidence clearly suggests, *then we have found the probable and ultimate origin and reason for why Native American "myths" of various ages of creatures and different "humanities" was so broadly accurate, for they did ultimately stem from an advanced culture in "high antiquity" that actually observed the "monsters" and wars described in their legends.*

Let us look closer at the chronological problem now posed by four entirely different data sets:

(1) the data set of modern genetics indicating the probable origin of modern man, *Homo sapiens sapiens*, approximately 150,000 years ago "out of Africa";
(2) the data presented by certain ancient texts as to the significant date of approximately 200,000 years ago;
(3) the data presented by geological and astronomical evidence that I recounted in *The Cosmic War;* and finally,

87 Ibid., pp. 120–122. Cremo and Thompson place a picture of the object on p. 121. I have also referred to it in my previous book, *The Cosmic War,* pp. 408–409. I survey Cremo's and Thompson's inventory of inconvenient artifacts on pp. 399–412 of that book. With respect to the curious South African machined "balls" I also note their peculiar resemblance to Saturn's apparently artificial moon, Iapetus (see *The Cosmic War,* p. 409).

(4) the anomalous and "inconvenient artifacts" presented by Cremo and Thompson.

Is there a way to *harmonize and synthesize* these very different models of human history? And if so, what *agenda or agendas* might that harmonization suggest was operative at different points of history and in different regions and cultures?

In my previous book *The Cosmic War,* I stated the following about the chronological problem, and the above list of disparate data sets:

> In order to tie together all the disparate pieces that I believe may form components of this gigantic scenario of cosmic war and catastrophe, of giants and chimeras, of "gods" and men and Nephilim, it is essential to paint in very broad strokes. While I do entertain discussion of broad chronological and other scientific and archaeological considerations, I do not enter into lengthy examinations of disciplines related to and affected by the Cosmic War hypothesis, such as evolutionary biology, anthropology, or even theology, philosophy, comparative religion, and esoteric or occult history. That such fields *are* affected by this hypothesis should be obvious. But to discuss each of these implications in detail would not only require several lengthy tomes in their own right, it would also distract attention from the main themes of the scenario...
>
> Similarly, I do not attempt to reconstruct a whole *detailed* chronology of an alternative "pre-history" of extraterrestrial contact, intervention, wars, and so on... for a very simple reason. The Cosmic War hypothesis has not hitherto been adequately advanced or explored in its own right, so it would seem best to ascertain its very broad outlines and progression and to put them forward here as a kind of *prima facie* case, and then to work out the detailed chronology at some later point.[88]

This is now that "later point," and as the focus in that book was the texts and *physics*, so the focus in this one is the texts and *biology and history.*

With these remarks in hand, and on the basis of the findings of the previous chapters of *this* book, we may lay out a more detailed chronology, commenting, as we proceed on the possible agendas at work.

88 Farrell, *The Cosmic War: Interplanetary Warfare, Modern Physics, and Ancient Texts,* pp. iii–iv.

C. Chronological Resolutions and Agendas: Some Speculations

1) Hundreds of millions, and perhaps even billions, of years before the advent of even the genus *Homo,* there were intelligent beings here on earth (and elsewhere in the solar system), who show evident signs of *technology and culture*. This is evident by two distinct data sets recounted here and elsewhere:
   a) *By the anomalous and inconvenient archaeological evidence assembled by Cremo and Thompson:*
      i) machined metallic balls found in South Africa in strata almost three billion years old;
      ii) a machined vase or cup found in Massachusetts in strata 600 million years old;
      iii) A nail embedded in Scottish sandstone between 360–408 million years old;
      iv) a gold thread embedded in rock between 320–360 million years old;
      v) a gold chain in coal 260–320 million years old;
      vi) an iron vase or cup in Oklahoma coal approximately 312 million years old;
      vii) a machined rectangular metallic tube found in chalk in France 65 million years old;

         This date, 65 million years ago, requires some commentary. In *The Cosmic War* I outlined the case that a planet once existed in the orbit of the asteroid belt, and that according to mathematical calculation, that planet had been exploded in a deliberate act of war, either some 65 million years ago, or 3.2 million years ago. While opting in that book for the 3.2 million year date for the event,[89] the presence of this artifact, and those chronologically preceding it, give some *prima facie* evidence for the existence of a civilization in high antiquity at least capable of machining things, and thus, the arguments advanced in *The Cosmic War* are broadly corroborated. The 65 million year date is significant for another reason, and that is that it is now an accepted model of science that some catastrophe occurred at that juncture of the Earth's planetary history that wiped out

89 Farrell, *The Cosmic War,* pp. 4–27.

the dinosaurs and ushered in the age of mammals.[90] It should thus be noted that *the presence of machined objects and evident "humanoid" culture on the Earth prior to the catastrophe of 65 million years ago, and which wiped out the dinosaurs, may be the ultimate basis for the later legends of wars against the giants and* ***monsters***. In other words, the later myths were *inherited* and based initially on *observation by someone of an event contemporaneous to them.* Given that we have argued the sophistication of this culture previously in this present book, and argued that there was a basis in ancient texts to assume that it had genetic engineering technology, and given the fact that the ancient texts examined in this present book have indicated the presence of chimeras and "monsters" — which monsters in other ancient texts were formed for the express purpose of weapons of war[91] — the possibility arises, radical as it may seem, that the dinosaurs themselves may have been engineered creatures, *since the artifact evidence of Cremo and Thompson indicates that intelligent beings were here from a period antedating the rise of the dinosaurs.*

viii). clay figurines found at Nampa, Idaho, two million years old; and last, but by no means least,

ix) a coin produced through a rolling mill, discovered in Illinois, between 200,000 and 400,000 years old.

This date also calls for some commentary. In the second chapter of this present book it was argued that the presence of geodetic and astronomical systems of measure that were advanced in Neolithic cultures indicated the presence of an elite with an agenda to make commerce a truly global affair. The existence of a coin *antedating* these efforts testifies to the existence of a civilization that had conducted such commerce and that for whatever reason had later to *re*-establish the basics necessary for it to resume: a system of accurate measures that would be reproducible world-wide. The coin, in other words, is minor, but corroborative, testimony to the existence of just such an elite with an agenda.

90 See the discussion in *The Cosmic War,* pp. 15–18.

91 See the discussion in *The Cosmic War,* pp. 150–166.

More significant, however, is the *date* of the coin, for it roughly correlates with, of all things, the dates given for the reign of the "divine" kings of Mesopotamia given in the Sumerian Kings' List. This list states explicitly that prior to the Deluge, these kings reigned for *241,200 years.*[92] Given that the Flood is generally assumed by some alternative researchers to have occurred approximately 10,000 B.C., we may add 12,000 years to that figure to get a rough approximate date of 253, 200 years to the arrival of these "divine" kings. The coin, in other words, is loose corroboration that the dates of the Kings' List, far from being the Sumerian fantasy that most scholars assume it to be, *might actually have some basis in fact.* Moreover, its presence in *Illinois* might be an indicator that far from being mere city-states, the cities mentioned in the Kings' List *might* have been the seats or capitals of very *large* empires spread over a wide surface of the globe.

b) *By the evidence assembled by Richard C. Hoagland that indicate a civilization sophisticated enough to be interplanetary once inhabited our own local solar system:*

i) At the head of this list must be the Saturnian moon, Iapetus, which, in addition to showing clear signs of being an artificial body, is also in an orbit *around* Saturn that incorporates Sumerian systems of measure;[93]

This also requires some additional commentary, for the presence of such measures in Iapetus' orbit around Saturn means that the Sumerian system of measures is incomparably *older* than Sumeria. It means likewise that the geodetic and astronomical bases of measures were known long before the Neolithic age, and thus suggest that in the aftermath of the Cosmic War, whensoever one dates it, enough survivors were left to perpetuate that knowledge and eventually reconstruct such systems of measure.

ii) There is evidence of the artificiality of at least one *asteroid*

---

92 For the complete text of the Sumerian Kings' List and my conclusions from it, see *The Cosmic War,* pp. 192–203.

93 Richard C. Hoagland, "A Moon With a View," www.enterprisemission.com, Part Six. The entire paper should be consulted for a thorough understanding of Mr. Hoagland's arguments. For a synopsized version, see my *The Cosmic War,* pp. 390–398.

in the asteroid belt,[94] and for the artificiality of the Martian moon Phobos;[95]

iii) There is evidence of the artificiality of structures and features on the surfaces of Mars[96] and Earth's own Moon;[97]

Again, these facts also require additional comment. Given that the standard method of dating the age of such objects is by crater counting, artificial objects such as Iapetus and Phobos, and the artificial structures on the Martian and Selene surfaces exhibit an antiquity approximately in line with the anomalous and inconvenient — and perhaps unearthly — artifacts unearthed by Cremo and Thompson. The *extraterrestrial archaeology is thus broadly in agreement with the terrestrial.*

At this juncture, we now add into the chronological table the previous "standard model" of the evolutionary origins of man; we repeat verbatim, in some instances, what was stated previously, with additional commentary:

2) Gorillas and chimpanzees split from the line leading to humans four to eight million years ago;
3) Australopithecines diverge from the genus *Homo* approximately 2.3–2.4 million years ago, that is — if the 3.2 million year date is accepted for the explosion of the missing planet of the solar system — *after* the occurrence of the Cosmic War;
4) *Homo habilis* flourished approximately 2.4–1.4 million years ago, evolving in southern and eastern Africa, diverging from Australopithecines;[98]
5) The next major evolutionary leap occurred with *Homo erectus*, who lived from approximately 1.8 million to a mere 70,000 years ago, the latter date indicating some contemporaneous existence with modern *Homo sapiens sapiens*. *Homo erectus* is believed also to have evolved larger brain capacity in some populations, and to fabricate and make use of simple stone tools, leading some to

94 Richard C. Hoagland, "Like a Diamond in the Sky...", www.enterprisemission.com.

95 Richard C. Hoagland, "For the World is Hollow and I Have Touched the Sky," www.enterprisemission.com.

96 Richard C. Hoagland, *The Monuments of Mars: A City on the Edge of Forever* (Berkeley, California: Frog Ltd., 2001), especially pp. 87–198.

97 Richard C. Hoagland and Mike Bara, *Dark Mission: The Secret History of NASA* (Port Townsend, Washington: Feral House, 2007), especially pp. 113–198.

98 Ibid., p. 5.

classify them as a separate species;[99]

Again we pause for commentary.

Given that it is possible to construe some of the texts examined in this book as indicative of a technology of genetic engineering in high antiquity, and given that the express purpose for the engineering of mankind by "the gods" was for the express purpose of relieving the burden of labor on the latter, we speculate that *in the wake of the Cosmic War, it was necessary to do so to insure the survival of that pre-existing civilization.* It will be recalled from chapter five's survey of the O'Briens' examination of the Kharsag Tablets that this engineering occurred by taking a pre-existing "humanid" and injecting it with a "divine" element. The possibility exists, in other words, when all the previous evidence of point number one above is considered, that the various "humanities" of the standard model might themselves be the products of engineering, or at least of a technological "assist" to normal evolutionary processes.

6) There is still debate over where *Homo neanderthalensis,* which lived from approximately *400,000 years ago* (a significant date as we shall see), is a part of the evolutionary tree of modern man or not; that is, is it a separate species or is it a sub-species of *Homo sapiens*? DNA evidence suggests that the two species shared a common ancestor no later than 660,000 years ago. The Wikipedia article notes, however, that "a recent development in 2010 indicates that Neanderthal did indeed interbreed with Homo Sapiens at cerca [sic] 75,000 B.C. to create modern humans..."[100]giving a DNA content of modern humans that is approximately one to four percent Neanderthal, a significant amount given that humans and chimpanzees differ only in 1.5 percent of DNA.[101] What is *most* interesting, however, is that this one to four percent of DNA that is common to modern man and Neanderthal man is present only in non-African humans;[102]
7) Sometime between 400,000 and 200,000 years ago differences in skull cranial capacity developed, accompanied with a similar development in the sophistication of stone tools, allowing paleontologists to speculate these populations are the first beginnings of the genus *Homo sapiens;*[103]

---

99 Ibid., p. 6.

100 "Human Evolution," *Wikipedia,* www.en.wikipedia.org/wiki/Human_evolution, p. 7.

101 Ibid.

102 Ibid.

103 Ibid., pp. 7–8.

8) Finally, as we have seen, the emergence of *Homo sapiens sapiens* occurred approximately 150,000 years ago, spreading from Africa, through the Middle East, to the rest of the world.

Points 6–8 above suggest that further genetic projects might have taken place. We have already pointed out the odd, though minor, corroboration of modern genetic findings with its 13 primordial African clans, and the Kharsag Tablets' 14 human donor mothers, and seven European clan mothers, to the genetic engineering project.

Oddly enough, as has been seen in the previous chapter, Native American traditions contain the idea of different ages populated by different "humanities," and in particular, as was also seen, the Aztec tradition recorded *five* different humanities, including the present one, *each of which* were engineered by "the gods." An ocean and hemisphere away, the Mesopotamian texts say essentially the same thing.

There is a further indicator that we might be looking at something of this nature, and that is the Sumerian tradition that "kingship" was "lowered from above" or from heaven, that it came from "the gods." When compared to the Kharsag Tablets it will be recalled that the "*divine*" part of modern came from "the gods" or, to be more accurate, one specific male donor, whereas the "human" component came from 14 donor *mothers.*

This odd combination might in fact reflect a dim mythological memory of two scientific facts we encountered in the previous chapter, namely that one can only trace *human* ancestry through the *mother* and mitochondrial DNA, whereas, if there were to be any traceable presence of DNA *unrelated* to humanity and coming from some other species of the genus *Homo*, then the most likely place to trace or discover it would be via the male Y chromosome which is found *only* in human *males.*[104] It is in fact within the male Y chromosome that there is a great deal of the so-called "junk" or non-coding DNA.

We recall also the fact of the more-than-coincidental resemblance of the Chinese *I Ching* system of divination to the nature of human DNA itself, a fact that strongly suggests that someone in ancient times knew a great deal not only about human DNA, but about its possible connection to a deeper physics. This suggests that while some texts and myths might have been garbled, preserving only faint traces of a lost plateau of scientific development, *other* traditions existed that preserved whole portions of that science intact — albeit encoded — and that in turn suggests the presence of an elite doing the preserving.

All of these things taken together suggest one other significant thing: an elite that may in fact be only *partially* human, and descended more directly

104 And in rare cases of human females, as was seen in the previous chapter.

from that putative genetic engineering of humanity, or perhaps even from the original "gods" themselves, an elite that perhaps has survived down to our own day, walking amongst us, undetected, "like aliens."

And this brings up the final point in our chronological table, and a final agenda:

9) At a *much* later point, some myths and texts arise whose function appears to be precisely to *obscure* the above scenario, and the elites behind the activities that comprise it. There is, so to speak, a *counter-revolution* underway by an elite opposing all these agendas and seeking to usurp power for itself. Its standard technique of obscurantism appears to be the "religification" of these ideas and a demand for unquestioning obedience to this or that dogma of this or that "god" and a willingness to murder under the justification of "obedience" to him.

As indicated in the Preface, I am acutely aware of the radical nature that this implication of the data *surveyed* here has for conventional religious apologetics, and for the history and possible editing of religious texts to suit an agenda. At the minimum, however, it is a growing problem, and serious scholars of texts and informed theologians, philosophers, anthropologists and scientists from every discipline will inevitably have to deal with it — and more successfully than the manner in which the past two millennia have dealt with it.

The time of idle and fruitless contemporary debates between the dialectical oppositions of the god of "intelligent design" versus the atheistic god of "evolutionary chance" are over, for if the synthetic view be true, then the truth lies somewhere between those polarities. The data from space, the data from texts, the moral and ethical dilemmas they pose, and the data from mankind and Earth herself, combine on this synthetic view into a scenario of vast and cosmic proportions, and the fate of humanity itself is at stake.

It is high time we get on with it, and to the work that this survey has merely scratched the surface of, for if any truths emerge from this survey they are these: (1) there is more truth than myth to ancient legends and texts, if examined a certain way, and (2) it is time for humanity to grow up, for after all, we are all, in the very real biological sense, cousins, and maybe, just maybe, we have other "cousins" out there somewhere still.

The bad news is, they may be coming back...

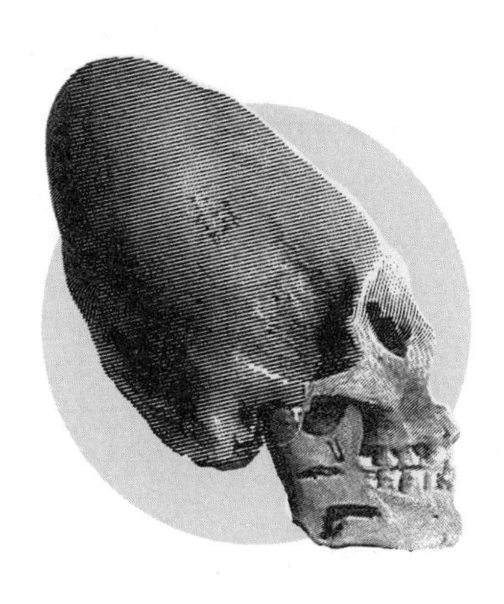

BIBLIOGRAPHY

Barton, George Aaron. *Miscellaneous Babylonian Inscriptions.* New Haven: Yale University Press, 1918. (Kessinger Publishing Reprint, ISBN 1120646510).

Bramley, William. *The Gods of Eden.* New York: Avon, 1990. ISBN 0-380-71807-3.

Brodeur, Paul. *The Zapping of America: Microwaves, Their Deadly Risk, and the Cover-Up.* New York: W.W. Norton and Company, Inc., 1977. ISBN 0-393-06427-1.

Cannon, Martin. *The Controllers: A New Hypothesis of Alien Abduction.* www.constitution.org/abus/controll.htm

Childress, David Hatcher. "Archeological Coverups." *Nexus,* 2005. www.unexplainable.net/artman/publish/article_1727.shtml

Childress, David Hatcher. "The Egyptian City of the Grand Canyon." *World Explorer Magazine.* Adventures Unlimited Press. 21–40.

Coppens, Philip. "Canyonitis: Seeing Evidence of Ancient Egypt in the Grand Canyon." www.philipcoppens.com/egyptiancanyon.html

Cremo, Michael A., and Richard L. Thompson. *Forbidden Archeology: The Hidden History of the Human Race.* Los Angeles: Bhaktivedanta Book Publishing, Inc., 1996. ISBN 0-89213-294-9.

Cremo, Michael A., and Richard L. Thompson. *The Hidden History of the Human Race (The Condensed Edition of* Forbidden Archeology). Angeles: Bhaktivedanta Book Publishing, Inc., 1999. ISBN 0-89213-325-2.

Delgado, Jose M.R., M.D. *Physical Control of the Mind: Toward a Psychocivilizaed Society.* New York: Harper Colophon Books, 1971. ISBN 06-090208-6.

Farrell, Joseph P. *Babylon's Banksters: The Alchemy of Deep Physics, High Finance, and Ancient Religion.* Port Townsend, Washington: Feral House, 2010. ISBN 978-1-932595-79-6.

Farrell, Joseph P. *The Cosmic War: Interplanetary Warfare, Modern Physics, and Ancient Texts.* Kempton, Illinois: Adventures Unlimited Press, 2007. ISBN 978-1-931882-75-0.

Farrell, Joseph P. *The Philosophers' Stone: Alchemy and the Secret Research for Exotic Matter.* Post Townsend, Washington: Feral House, 2009. ISBN 978-1-032595-40-6.

Fort, Charles. *The Book of the Damned* from *The Collected Works of Charles Fort.* New York: Tarcher-Penguin, 2008. ISBN 978-58542-641-6.

Hayes, Michael. *The Hermetic Code in DNA: The Sacred Principles in the Ordering of the Universe.* Rochester, Vermont: Inner Traditions, 2008. ISBN 978-159477218-4.

Hoagland, Richard C., and Mike Bara. *Dark Mission: The Secret History of NASA.* Port Townsend, Washington: Feral House, 2007. ISBN 978-932595-26-0.

Knight, Christopher, and Alan Butler. *Before the Pyramids: Cracking Archaeology's Greatest Mystery.* London: Watkins Publishing, 2009. ISBN 978-1-906787-38-7.

Knight, Christopher, and Alan Butler. *Civilization One: The World is Not As You Thought It Was.* London: Watkins Publishing, 2004. ISBN 1-84293-095-8.

Mayor, Adrienne. *Fossil Legends of the First Americans.* Princeton: Princeton University Press, 2005. ISBN 978-0-691-13049-1.

Mayor, Adrienne. *Greek Fire, Poison Arrows, and Scorpion Bombs: Biological and Chemical Warfare in the Ancient World.* New York: Overlook Duckworth, 2009. ISBN 978-1-59020-177-0.

Mayor, Adrienne. *The First Fossil Hunters: Paleontology in Greek and Roman Times.* Princeton: Princeton University Press, 2000. ISBN 0-691-08977-9.

Meyl, Konstantin. *Scalar Waves: From an Extended Vortex and Field Theory to a Technical, Biological, and Historical Use of Longitudinal Waves.* Villingen-Schwenningen, 2003. ISBN 3-9802-542-4-0.

O'Brien, Christian, and Barbara Joy O'Brien. *The Genius of the Few: The Story of Those Who Founded the Garden in Eden.* Dianthus Publishing Limited, 1999. ISBN 0-946604-17-7.

Ridley, Matt. *Genome: The Autobiography of a Species in 23 Chapters.* New York: Harper Perennial. ISBN 978-0-06-089408-5.

Schönberger, Dr. Martin. *The I Ching and the Genetic Code: The Hidden Key to Life.* Santa Fe, New Mexico: Aurora Press, 1992. ISBN 0-943358-37-X.

Shreeve, James. *The Genome War: How Craig Venter Tried to Capture the Code of Life and Save the World.* New York: Ballantine Books. ISBN 978-0-345-43374-9.

Sitchin, Zechariah. *There Were Giants Upon the Earth: Gods, Demigods, and Human Ancestry: The Evidence of Alien DNA.* Rochester, Vermont: Bear and Company, 2010. ISBN 978-159143121-3.

Sykes, Bryan. *The Seven Daughters of Eve: The Science That Reveals our Genetic Ancestry.* New York: W.W. Norton & Company, 2001. ISBN 978-0-393-3214-6.

OTHER FERAL HOUSE TITLES BY JOSEPH FARRELL

BABYLON'S BANKSTERS

The Alchemy of Deep Physics, High Finance and Ancient Religion